Biological Test Method: Reference Method for Determining Acute Lethality of Effluents to Rainbow Trout

Method Development and Applications Section
Environmental Technology Centre
Environment Canada
Ottawa, Ontario

Report EPS 1/RM/13
SECOND EDITION
December 2000

National Library of Canada Cataloguing in Publication Data

Main entry under title:

Biological test method. Reference method for determining acute lethality of effluents to rainbow trout

Second ed.
(Report EPS 1/RM/13)

Text in English and French on inverted pages
Title on added t.p.: Méthode de référence pour la détermination de la léthalité aiguë chez la truite arc-en-ciel.
Includes bibliographical references.
ISBN 0-660-61615-7
Cat. no. En49-24/1-13-2001

1. Rainbow trout -- Mortality -- Testing -- Standards -- Canada.
2. Rainbow trout -- Microbiology -- Standards -- Canada.
3. Effluent quality -- Testing -- Standards -- Canada.
4. Toxicology – Methodology.

I. Environmental Technology Centre (Canada), Method Development and Applications Section.
II. Canada. Environment Canada.
III. Series: Report (Canada. Environment Canada); EPS 1/RM/13.

QH90.57B56 2001 597.5'7 C2001-980086-XE

Catalogue No. En 49-24/1-13-2001
ISBN 0-660-61615-7

Readers' Comments

Comments regarding the content of this report should be addressed to:

Richard Scroggins
Method Development and Applications Section
Environmental Technology Centre
Environment Canada
3439 River Road South
Ottawa, Ontario
K1A 0H3

Review Notice

This report has been reviewed by the staff of the Environmental Technology Advancement Directorate, Environment Canada, and approved for publication. Mention of trade names or commercial products does not constitute endorsement by Environment Canada. Other products of similar value are available.

Abstract

Explicit standard or reference methods for measuring the acute lethal toxicity of effluents to rainbow trout (Oncorhynchus mykiss) *are provided in this report. Specific instructions for performing acute lethality tests with samples of effluent are given. The guidance provided in the generic methodology report "Acute Lethality Test Using Rainbow Trout" (EC, 1990a, including its May 1996 Amendments) is built upon herein.*

The present report represents the second edition of Reference Method EPS 1/RM/13, published in July 1990 and amended in May 1996 (EC, 1990b, including its May 1996 Amendments). It supersedes that earlier version, and is to be applied as Environment Canada's current reference method *for determining the acute lethality of effluents to rainbow trout.*

Methods are given for: 1) a single-concentration test, with full-strength effluent unless otherwise specified; 2) a multi-concentration test to determine the median lethal concentration (LC50); and 3) a test with a reference toxicant. Instructions are included on holding trout in the laboratory, facilities and water supply, handling and storage of samples, preparation of solutions, test conditions, observations to be made, endpoints with methods of calculation, and the use of reference toxicants.

Foreword

This is one of a series of **reference methods** *for measuring and assessing the toxic effect(s) on single species of aquatic or terrestrial organisms, caused by their exposure to samples of test materials or substances under controlled and defined laboratory conditions.*

A **reference method** *is defined herein as a specific biological test method for performing a toxicity test, i.e., a toxicity test method with an explicit set of instructions and conditions which are described precisely in a written document. Unlike other multi-purpose (generic) biological test methods published by Environment Canada, the use of a* **reference method** *is frequently restricted to testing requirements associated with specific regulations.*

Reference methods *are those that have been developed and published by Environment Canada (EC), and are favoured:*

- *for regulatory use in the environmental toxicity laboratories of federal and provincial agencies;*
- *for regulatory testing which is contracted out by Environment Canada or requested from outside agencies or industry;*
- *for incorporation in federal, provincial, or municipal environmental regulations or permits, as a regulatory monitoring requirement; and*
- *as a foundation for the provision of very explicit instructions.*

Table of Contents

Abstract . . . . **v**
Foreword . . . . **vi**
Terminology . . . . **ix**
Acknowledgements . . . . **xii**

Section 1
Introduction 1

Section 2
Organisms and Holding 3
2.1 Species and Source 3
2.2 Holding and Acclimation 3
2.3 Water 4
2.4 Physicochemical Conditions 5

Section 3
Facilities 6

Section 4
General Procedure for Determining Acute Lethality of Effluent 7
4.1 Sample Labelling, Transport, and Storage 7
4.2 Test Conditions 7
4.3 Preparing Test Solutions 8
4.4 Beginning the Test 9
4.5 Observations and Measurements 9

Section 5
Procedure for a Single-concentration Test
to Determine the Mortality Rate at 96 Hours 11

Section 6
Procedure for a Multi-concentration Test to Determine
the 96-h LC50 12

Section 7
Procedure for Tests with a Reference Toxicant 13

Section 8
Reporting Requirements 15
8.1 Data to be Reported 15
8.1.1 Effluent 15

8.1.2 Test Facilities and Conditions 15
8.1.3 Results 16
8.2 Data to be Held on File 16
8.2.1 Effluent 16
8.2.2 Test Facilities and Conditions 17
8.2.3 Results 18

References 19

Appendix
Members of the Inter-Governmental Aquatic Toxicity Group and Addresses of Environment Canada's Headquarters and Regional Offices 21

Terminology

The following definitions are given in the context of the procedures in this report. Additional definitions in the detailed companion document (EC, 1990a, including its May 1996 Amendments) apply here.

Grammatical Terms

Must is used to express an absolute requirement.
Should is used to state that the specified condition or procedure is recommended and ought to be met if possible.
May is used to mean "is (are) allowed to".
Can is used to mean "is (are) able to".

Technical Terms

Acclimation means to become physiologically adapted to a particular level of one or more environmental variables such as temperature. The term usually refers to controlled laboratory conditions.

Acute means happening within a short period of time, usually taken as ≤96 h for fish.

Alevin is a recently-hatched, non-feeding fish with an evident yolk sac (for nutritive requirements). It is often referred to as yolk-sac fry.

Conductivity is a numerical expression of the ability of an aqueous solution to carry an electric current. This ability depends on concentrations of ions in solution, their valence and mobility, and on temperature. Conductivity is normally reported as millisiemens/metre, an SI unit (Système internationales d'unités), or as micromhos/cm (1 mS/m = 10 µmhos/cm).

Control is a treatment in an investigation or study that duplicates all the conditions and factors that might affect the results, except the specific condition that is being studied. In an aquatic toxicity test, the control must duplicate all conditions of the exposure treatment(s), but must contain no test material. The control is used to determine the absence of measurable toxicity due to basic test conditions (e.g., temperature, quality of dilution water, health of test organisms, or effects due to their handling).

Control/dilution water is the water used for diluting the sample of effluent, and for the control test.

Dechlorinated water is a chlorinated water (usually municipal drinking water) that has been treated to remove chlorine and chlorinated compounds from solution.

Dilution water is that which is used to dilute a test material, to prepare different concentrations for the toxicity test.

Effluent is any liquid waste (e.g., industrial, municipal) discharged to the aquatic environment.

Fingerling is a young (underyearling), actively feeding fish.

Flow-through describes tests in which solutions in test vessels are renewed continuously by the constant inflow of a fresh solution, or by a frequent intermittent inflow.

Fork Length is the length of a fish, measured from the tip of the nose to the fork of the tail.

Hardness is the concentration of cations in water that will react with a sodium soap to precipitate an insoluble residue. In general, hardness is a measure of the concentration of calcium and magnesium ions in water, expressed as mg/L calcium carbonate.

LC50 (median lethal concentration) is the concentration of material (in this case, effluent) in water that is estimated to be lethal to 50% of the test organisms. The LC50 and its 95% confidence limits are usually derived by statistical analysis of percent mortalities in several test concentrations, after a fixed period of exposure. The duration of exposure must be specified (e.g., 96-h LC50).

Lethal means causing death by direct action. Death of fish is defined here as the cessation of all visible signs of movement or other activity.

Lux is a unit of illumination based on units per square metre. One lux = 0.0929 foot-candles and one foot-candle = 10.76 lux.

Overt means obviously discernible under the test conditions employed.

pH is the negative logarithm of the activity of hydrogen ions in gram equivalents per litre. The pH value expresses the degree or intensity of both acidic and alkaline reactions on a scale from 0 to 14, with 7 representing neutrality, numbers less then 7 signifying increasingly greater acidic reactions, and numbers greater than 7 indicating increasingly basic or alkaline reactions.

Photoperiod is the duration of illumination and darkness within a 24-h day.

Pretreatment means, in this report, treatment of a sample or dilution thereof, prior to exposure of fish.

Reference method refers to a specific biological test method for performing a toxicity test, i.e., a toxicity test method with an explicit set of instructions and conditions which are described precisely in a written document. Unlike other multi-purpose (generic) biological test methods published by Environment Canada, the use of a *reference method* is frequently restricted to testing requirements associated with specific regulations; testing to assess whether there has been a violation of the General Provisions of the Canadian Fisheries Act.

Reference toxicant is a standard chemical used to measure the sensitivity of the test organisms to establish confidence in the toxicity data obtained for a test material. In most instances, a toxicity test with a reference toxicant is performed to assess the sensitivity of the organisms at the time the test material is evaluated, and the precision of results obtained by the laboratory for that chemical.

Salinity is the total amount of solid substance, in grams, dissolved in 1 kg of water. It is determined after all carbonates have been converted to oxides, all bromide and iodide have been replaced by chloride, and all organic matter has been oxidized. Salinity can also be measured directly using a salinity/conductivity meter or other means (see APHA *et al.*, 1998). It is usually reported in grams per kilogram or parts per thousand (‰).

Static describes toxicity tests in which test solutions are not renewed during the test.

Static replacement describes toxicity tests in which test solutions are renewed (replaced) periodically during the test, usually every 24 h. Synonymous terms are "renewal", "batch replacement", and "semi-static".

Sublethal means detrimental to the fish, but below the level which directly causes death within the test period.

Swim-up fry is a young, post-alevin fish which has commenced active feeding.

Toxicity is the inherent potential or capacity of a material to cause adverse effect(s) on fish or other living organisms. The effect(s) could be lethal or sublethal.

Acknowledgements

The first edition of this report, which was printed in July 1990 and amended in May 1996 (EC, 1990b), was co-authored by D.J. McLeay (McLeay Associates Ltd., West Vancouver, BC) and J.B. Sprague (Sprague Associates Ltd., Guelph, ON). Messrs. R. Scroggins and G. Sergy (EPS, Environment Canada) acted as Scientific Authorities and provided technical input and guidance. This first edition was shaped by the objectives, needs, and comments of members of a subcommittee from the Environmental Protection Service (EPS): N. Bermingham (EPS, Montreal); C. Blaise (EPS, Longueuil); G. Coulombe (EPS, Montreal); K. Doe (EPS, Dartmouth); G. Elliott (EPS, Edmonton); D. MacGregor (EPS, Ottawa); J. MacLatchy (EPS, Ottawa); W. Parker (EPS, Dartmouth); D. Robinson (Fisheries & Oceans Canada, Ottawa); T. Ruthman (EPS, Ottawa); R. Scroggins (EPS, Ottawa); G. Sergy (EPS, Edmonton); and R. Watts (EPS, North Vancouver). Provincial members of the Inter-Governmental Aquatic Toxicity Group active in 1990 (i.e., G. Joubert, ministère de l'Environnement du Québec; G. Westlake, Ontario Ministry of Environment; W. Young, Manitoba Environment and Public Safety; K. Lauten, Saskatchewan Environment and Public Safety; J. Somers, Alberta Environment; S. Horvath and G. van Aggelen, B.C. Ministry of Environment) provided valuable input and comment at that time. The following persons also reviewed drafts of the first edition and contributed to its publication: D. Vaughan (EPS, Dartmouth); S. Wade (EPS, Dartmouth); A. Beckett (EPS, Edmonton); F. Zaal (EPS, Edmonton); S. Yee (EPS, North Vancouver); D. Moul (EPS, North Vancouver); V. Zitko (Fisheries & Oceans Canada, St. Andrews); V. Cairns (Fisheries & Oceans Canada, Burlington); S. Samis (Fisheries & Oceans Canada, Vancouver); R. Martin (Newfoundland Dept. Environ. & Lands, St. John's); J. Lee (Ontario Ministry of Environment, Rexdale); J. Moores (Alberta Environment, Vegreville); P. Martel (Pulp & Paper Research Institute of Canada, Pointe Claire); Y. Bois (Technitrol · Eco Inc., Pointe Claire); C. R. Cook (E.B. Eddy Forest Products Ltd., Espanola); G. Craig (Beak Consultants Ltd., Brampton); and D. Bradley, D. Monteith, and J. Pickard (B.C. Research Corp., Vancouver). The laboratory testing support of Environment Canada (see Appendix) is also acknowledged.

The following persons contributed to the May 1996 Amendments to the first edition of this *reference method* (EC, 1990b), which are now incorporated (with changes, as deemed appropriate) in this second edition: C. Bastien (ministère de l'Environnement du Québec, Sainte-Foy); B. Bayer (Manitoba Environment, Winnipeg); K. Doe (Environmental Conservation Service [ECS], Moncton); G. Elliott (EPS, Edmonton); P. Jackman (ECS, St. John's); J. Miller (EPS, Ottawa); G. van Aggelen (ECS, North Vancouver); G. Westlake (Ontario Ministry of Environment, Rexdale); and R. Watts (ECS, North Vancouver).

This second edition was prepared by D.J. McLeay (McLeay Environmental Ltd., Victoria), with guidance and input from R.P. Scroggins (Method Development and Applications Section, Environment Canada, Ottawa). Members of the Inter-Governmental Environmental Toxicity Group (IGETG; see Appendix) contributed to and supported the modifications in this edition of the report. Special acknowledgement is made of the following persons, who prompted certain changes and additions: K. Doe (ECS, Moncton); G. Elliott (EPS, Edmonton); and J. Schroeder (Ontario Ministry of Environment, Rexdale).

Introduction

This report replaces the first edition of Environment Canada's *reference method* EPS 1/RM/13 (EC, 1990b), which was published in July 1990 and amended in May 1996. Testing instructions, guidance, and wording in this (second) edition of EPS 1/RM/13 remain the same as in the amended first edition except for the changes or additions included here, which were considered by Environment Canada to be necessary or prudent improvements.

The procedures for an acute lethality test with rainbow trout, as specified by Canadian governments involved in pollution monitoring and control of industrial or municipal effluents, are given in this report. Rainbow trout have been used for three decades in Canada for testing effluents under a series of regulations and guidelines (EPS, 1971, 1973, 1974, 1977a–c, 1980, 1984; Government of Canada, 1992). This *reference method* should be used in conjunction with a more extensive report which gives supporting rationale and additional details (EC, 1990a, including its 1996 Amendments).

Many components of procedures in this report are similar to Canadian provincial methods (McGuinness, 1982; Rocchini *et al.*, 1982; Craig *et al.*, 1983; OME, 1989), methods used in the United States (ASTM, 1980; USEPA, 1985a, 1985b; APHA *et al.*, 1998), or international techniques (BHSC, 1982; UKWRC, 1983; OECD, 1984). The contribution of the above-mentioned methods to all parts of the present report is acknowledged, and they are recommended as sources of supporting information. Procedures stipulated in this *reference method* should, however, be taken as the definitive ones for regulatory purposes.

Rainbow trout *(Oncorhynchus mykiss*; formerly *Salmo gairdneri*) are to be used as the test organisms when performing this *reference method.* This species is native to western North America. It now inhabits waters of all Canadian provinces and has been widely introduced around the world. It thrives in cool, fresh water, runs to the sea on both Atlantic and Pacific coasts, and is commonly reared in hatcheries and commercial aquaculture operations. It has also become the world's standard cool-water fish for freshwater toxicity tests, with a toxicological data bank of appreciable magnitude.

Three basic procedures are described. One uses a single concentration of effluent (full strength unless otherwise specified) and a control, as would be suitable for a pass/fail test. A second procedure estimates the median lethal concentration (LC50) (i.e., it determines the degree of toxicity using several concentrations of effluent including full strength). A third procedure is a multi-concentration test with a reference toxicant, to assess the sensitivity of the test fish to a standard toxicant and the precision of the data produced by the laboratory for that chemical.

This test is to be used with effluents containing fresh water or having a salinity of ≤10‰, defined as conductivity ≤1400 mS/m at a temperature of 15°C. Saline (>10‰) effluents discharging into fresh water should also be tested with rainbow trout acclimated to fresh water. Saline (>10‰) effluents discharging directly to estuarine or marine receiving waters should be tested with a species authorized by the regional Environment Canada laboratory (see Appendix) and acclimated to seawater of similar salinity to that of the effluent.

Organisms and Holding

2.1 *Species and Source*

Test fish are to be rainbow trout *(Oncorhynchus mykiss*); either swim-up fry that have been actively feeding for at least two weeks, or fingerlings. Their average wet weight must be between 0.3 and 2.5 g, and the largest fish should not be more than twice the length of the smallest in the same test.

Fish may be acquired as eyed eggs, fry, or fingerlings. Each group of fish should be obtained from a certified "disease-free" hatchery with an ongoing health monitoring and certification program. Procurement and shipment of fish should be approved by regional representatives of the Federal (Fisheries and Oceans Canada) – Provincial Transplant Committee, in provinces where this committee acts to control movements of fish stocks. Advice on sources of trout can be obtained from regional offices of Environment Canada (see Appendix).

2.2 *Holding and Acclimation*

Fish should be reared using tanks and other facilities made of nontoxic materials such as stainless steel, porcelain, fibreglass-reinforced polyester, acrylic, polyethylene, or polypropylene. Eggs and alevins may be incubated in vertical-flow hatchery trays or flowing water troughs (Leitritz and Lewis, 1976). Fry and fingerlings may be reared and acclimated in troughs or tanks with flowing water, located away from physical disturbances and preferably in a location separate from the test tanks.

Following the transport of fish to the rearing/acclimation facilities of the testing laboratory, they must be held under conditions specified in Section 2.4 for a minimum period of two weeks. This acclimation period must immediately precede their use in a test. Fish acclimation may be done indoors, or outdoors using lids with photoperiod-controlled lights. See Environment Canada (1990a) for additional details on holding and acclimating fish for use in toxicity tests.

Tanks should be kept clean, with siphoning of excess food and faeces as frequently as necessary. Tanks with central, double standpipes are partially self-cleaning and are recommended. Tanks should be disinfected and rinsed thoroughly with water used for holding/acclimating fish before introducing a new batch of fish. Disinfectants such as those containing chlorinated or iodophore compounds or n-alkyldimethyl-benzylammonium chloride should be used.

Unless specified otherwise by the feed manufacturer, feeding should be once or more per day with a recognized (standard) commercial pelleted fish food, at a daily ration approximating 1 to 5% of wet body weight, depending on temperature and fish size (EC, 1990a). Pellet size and type should be chosen in consideration of fish size and age, water temperature, and the manufacturer's recommendations. The duration and method of food storage should also follow the manufacturer's recommendation.

Dead or moribund fish should be removed immediately after daily inspection.

Mortalities in the stock tank(s) from which test fish are to be taken should be monitored and recorded daily, and as a minimum must be monitored and recorded five days per week. The cumulative rate of fish mortality during the 7-day period preceding the day that the toxicity test is started must be less than 2%. If the cumulative fish mortality rate during this period is 2 to 10%, acclimation must be extended for at least an additional seven days and until a cumulative 7-day mortality rate of <2% is achieved for the 7-day period preceding the day that the toxicity test is started. A cumulative mortality rate of >10% per week, during any 7-day period, makes the group of fish unacceptable for future use if deaths are caused by disease or aquatic contaminants. If deaths result from other factors (e.g., high initial mortalities during transition from alevins to swim-up fry or following fish transfer), the fish may be used for future toxicity tests provided that mortalities in the stock tank(s) from which fish are to be taken decline to <2% during the seven days immediately preceding the day that the test is started.

Chemical treatment of diseased fish should be avoided. If the use of chemically treated fish cannot be avoided, a minimum two-week period must follow their treatment before they are used in tests. The test with a reference toxicant (see Section 7) gives some indication of the suitability of the fish for use in toxicity tests.

2.3 Water

Water for holding and acclimating fish can be uncontaminated groundwater, surface water, or dechlorinated municipal drinking water. The water supply should consistently support good survival, health, and growth of trout. Chemical quality of the laboratory's water supply should be measured as often as necessary to document variation. This should include at least hardness, pH, conductivity, dissolved oxygen, and residual chlorine (if municipal drinking water is used), and as appropriate, alkalinity, suspended solids, total organic carbon, total dissolved gases, ammonia, nitrite, metals, and total organophosphorus pesticides. Any supersaturation with gases should be remedied (see EC, 1990a).

If dechlorinated municipal drinking water is used, it must be free of any harmful concentration of chlorine or chlorinated compounds upon fish exposure. The target value for total residual chlorine in holding tanks, and for that in control/dilution water within test tanks, is ≤0.002 mg/L (see EC, 1990a).

Flow of fresh (new) water through tanks used for holding and acclimating fish must be ≥1.0 L/min for every kilogram of fish being held (1.4 L/g fish · d or 0.69 g fish · d/L). In addition, a tank must contain at any given time, ≥1.0 L/10 g fish. Water flow rates to each tank should be measured and recorded at regular (e.g., daily or, as a minimum, weekly) intervals. Additionally, ten or more fish should be removed randomly from each holding/acclimation tank at regular (e.g., weekly) intervals to determine individual wet weights and assure that these requirements are met. The mean wet weight of individual fish should be determined and recorded, for each of these samples. These measurements should also be used as a guide when determining the volume of effluent required for a test, and to ensure that the maximum loading density of 0.5 g fish/L solution in each test vessel during the toxicity test (see Section 4.2) will not be exceeded.

2.4 Physicochemical Conditions

Lighting should be full spectrum, with 100 to 500 lux intensity at the water surface. For at least two weeks before a test, photoperiod must be constant at 16 ± 1 h of light and 8 ± 1 h of darkness, preferably with a 15- to 30-minute transition period.

Holding temperature may be 4 to 18°C, but fish must be acclimated for ≥2 weeks, and preferably ≥3 weeks, at 15 ± 2°C before use in a test. Change between temperature levels may proceed at ≤3°C/d. Dissolved oxygen within tanks should be 80 to 100% air saturation. Supplementary aeration to the tanks should be provided if necessary, using filtered, oil-free compressed air. The pH of water should be within the range of 6.0 to 8.5. Temperature, oxygen, pH, flow, and fish mortalities should be monitored for each holding or acclimation tank, preferably daily; weekly or more frequent monitoring of levels of ammonia and nitrite in holding or acclimation tanks is recommended. Weekly or more frequent monitoring of water in these tanks for total residual chlorine is also recommended if dechlorinated municipal drinking water is used as the water source.

Facilities

Tests must be performed in a facility isolated from general laboratory disturbances, either a separate room or a section walled or curtained off. Dust and fumes should be minimized. Control of test temperature (15 ± 1 °C) may be achieved by thermostatically controlled air conditioning or by immersing test vessels in regulated water baths.

Test vessels and all other equipment that may contact the test solutions or control/dilution water must not contain leachable substances, nor should they sorb toxicants from the test solution. Test vessels must be glass or Plexiglas™, acrylic, polypropylene, polyethylene, or have high quality (nontoxic) polyethylene liners. If liners are used, they must be discarded at the end of the test. It is recommended that test vessels be covered if necessary to prevent fish from escaping (EC, 1990a). All containers used for a test must be identical, and the minimum water depth must be 15 cm. Equipment must be thoroughly cleaned and rinsed in accordance with good laboratory procedures.

The control/dilution water should be the type described in Section 2.3, and it should preferably be identical to that used for holding and/or acclimating the fish.

General Procedure for Determining Acute Lethality of Effluent

4.1 Sample Labelling, Transport, and Storage

Sample-volume requirements depend on fish size and numbers per test solution, loading-density requirements, test concentrations, and the use of replicates. For single-concentration tests, sample volumes of 25 to 50 L or more are normally required. For tests to determine an LC50, sample volumes of 50 to 100 L or more are normally required.

Containers for storage and transportation of samples must be made of nontoxic material (e.g., polyethylene or polypropylene carboys or pails). The containers must be new or thoroughly cleaned and dried, and should be rinsed with clean water, then with the sample to be collected. They should be filled with sample to exclude air and then sealed (e.g., using a snap-on lid if the sample container is a pail). Labelling must include at least sample type, source, date and time of collection, and name of sampler(s).

Samples must be kept from freezing. During transport, samples should be kept in the dark, and at a temperature of 1 to 8° C if more than two days are spent in transit. Upon the receipt of sample(s) at the laboratory, the temperature of the effluent in each sample container should be measured and recorded. That portion of each sample to be used in the toxicity test must be adjusted to 15 ± 1 °C before the toxicity test can be started.

To enable the toxicity test to be started on the day that the sample is received in the laboratory, temperature adjustment of the effluent sample(s) may be done quickly (see Section 4.3). Alternatively, the laboratory may choose to store the sample in the dark at 4 ± 2 °C for a brief period (e.g., over the weekend, if the sample arrived on a Friday afternoon). Using this option, the sample must be stored in full, sealed containers which are held in the dark within a refrigerated facility. A third option is to hold the sample overnight within a facility adjusted to the test temperature (i.e., 15 ± 1 °C), in which instance the test must be started the next day. If a sample is warmed or cooled at 15 ± 1 °C overnight, it must be kept in one or more full, sealed containers during that time.

Testing of samples should commence as soon as possible after collection. The test should begin within three days and must commence no later than five days after termination of sampling. The contents of each sample container must be agitated thoroughly just before pouring aliquots to prepare solutions. Sub-samples (i.e., aliquots of a sample divided between two or more containers) must be combined.

4.2 Test Conditions

This is a 96-h static test, i.e., there is no replacement of solutions. Loading of fish into each test vessel must not exceed a density of 0.5 g/L; adherence to this requirement is based on the mean wet weight of control fish at the end of the test (Section 4.5). Fish must not be fed during the test, nor during the 16-h period immediately preceding it. The test is not valid if >10% of control fish die or exhibit atypical/stressed behaviour (EC, 1990a).

The test must be conducted at 15 ± 1 °C. All solutions must be aerated throughout the test, at a controlled rate of 6.5 ± 1 mL/min · L. Lighting must be the same as that defined for acclimation (see Section 2.4). Photoperiod (a light : dark cycle of 16 ± 1 h : 8 ± 1 h) must coincide with the timing which prevailed during acclimation.

The test must be conducted without adjustment of sample or solution pH. However, if it is desired to understand the extent to which extremes in solution or sample pH (e.g., outside the range of 5.5 to 8.5) may contribute to acute lethality, a parallel (pH-adjusted) test may be used. If both pH-adjusted and non-adjusted tests are run, definitive results should be those derived from the non-adjusted test. Rationale and procedural details regarding pH adjustment are provided in Environment Canada (1990a). Adjustment of pH is also one of a number of "Toxicity Identification Evaluation" techniques for characterizing the cause of sample toxicity (USEPA, 1991).

4.3 Preparing Test Solutions

Adjustment of the effluent sample and control/dilution water to 15 ± 1 °C must be done if the temperature is outside that range. The sample may be cooled using a cold-water bath or immersion cooler made of nontoxic material, or warmed using a hot-water bath. Samples or test solutions must not be heated by immersion heaters.

For a given test, the same water is to be used for preparing the control(s) and all test concentrations less than 100%. This is almost always the same water as used for acclimation. If the temperature of this water is adjusted upwards, supersaturation with gases must be avoided. The water must have an oxygen content within the range of 90 to 100% air saturation, achieved if necessary by vigorous aeration with oil-free compressed air passed through air stones or glass diffusers.

Test vessels should be rinsed with control/dilution water just before use, although that is not necessary if disposable polyethylene liners are used. Each test solution must be made up to an identical volume, and well mixed with a glass rod, Teflon™ stir bar, or other device made of nontoxic material, just before its use. All test vessels, measurement devices, stirring equipment, and fish-transfer pails must be thoroughly cleaned and rinsed in accordance with standard operational procedures.

The depth of solution in each test vessel must be at least 15 cm. Upon preparation of the test solutions, each must be aerated for a period of 30 minutes at 6.5 ± 1 mL/min · L. Thereafter, the concentration of dissolved oxygen must be measured in at least the highest test concentration (normally 100% effluent). If (and only if) oxygen in the highest test concentration is <70% or >100% of air saturation, then pre-aeration (i.e., before exposure of fish) of all solutions including the control(s) must be continued at 6.5 ± 1 mL/min · L. This period of pre-aeration must be restricted to the lesser of 90 additional minutes and attaining 70% saturation in the highest test concentration (or 100% saturation if super-saturation is evident). Immediately thereafter, fish must be placed in each test solution and the test initiated, regardless of whether 70 to 100% saturation was achieved in all test solutions. Aeration of test solutions should be provided by bubbling compressed air through a clean silica-glass air diffuser or disposable glass pipette (EC, 1990a). Aeration of each test solution at a rate of 6.5 ± 1 mL/min · L must be continued throughout the test.

4.4 Beginning the Test

One or more control solutions must be prepared and included as part of each test conducted on each sample. The multiple use of a control solution and its fish for more than one toxicity test and/or more than one effluent sample is unacceptable.

Each test vessel must be clearly coded or labelled as to concentration, date, and time of start. Vessels should be positioned for easy observation of fish. If a multi-concentration test is being performed (Section 6), the concentrations should be positioned at random.

Healthy fish which have been acclimated for a minimum of two weeks to the temperature and lighting conditions used in the test (see Sections 2.1, 2.2, 2.3, and 2.4) must be taken randomly from the acclimation tank(s) for use in the test. Handling and transfer procedures should minimize stress (EC, 1990a).

At least ten fish must be introduced into each test concentration including each control solution. They may be divided between two or more vessels at the same concentration to meet the required limit on loading (see Section 4.2).

Besides positioning the test vessels at random within the testing facility, the order of adding fish to each test solution should also be randomized. One or two fish should be introduced sequentially to each test solution, including the control solution(s), until 10 fish have been placed in each.

If one or more test solutions are highly coloured, opaque, or foamy, baskets made of nontoxic, nonabrasive material (e.g., nylon, polyethylene, polypropylene) may be used to permit inspections of fish during the test. If used, a basket should be placed in each test vessel including the control(s). Baskets should be big enough to allow fish to move throughout the test vessel. Each basket must be thoroughly cleaned and rinsed with control/dilution water before being used.

4.5 Observations and Measurements

Colour, turbidity, odour, and floating or settling solids in the sample should be noted at the start of the test. The appearance of test solutions should also be noted, and any obvious changes during the test should be recorded.

Measurements of dissolved oxygen, pH, and temperature must be made in each test solution including the control(s), at the start and end of the test as a minimum. Final measurements should be done after biological observations are complete. Conductivity of each test solution must be measured at the start of the test as a minimum.

The frequent and routine observation of fish in each test vessel is required to obtain information regarding their time to death as well as any overt sublethal responses evident, and to remove dead fish which could otherwise foul the test solution. Whenever possible, fish in each test vessel should be inspected at least at 24, 48, 72, and 96 h;

more frequent observations are recommended during the initial day of the test. During each observation period, all dead fish must be recorded and removed. Fish are considered to be dead when they fail to show evidence of opercular or other activity, and do not respond to subsequent gentle prodding. Overt sublethal toxic effects should also be recorded (see EC, 1990a). For highly coloured, opaque, or foamy test solutions, fish may be inspected using a dip net (cleaned and rinsed before use) or by raising them to the surface within a suitable basket (Section 4.4).

The mean fork length of control fish must be determined and recorded at the end of the test. The measurements of fork length, for each control fish, should be used to assess if the size of the test fish met with the recommendation that "*the largest fish should not be more than twice the length of the smallest in the same test*" (Section 2.1).

The mean wet weight of individual control fish must be determined and recorded at the end of the test. This measurement must be used to confirm that the acceptable range of weights for test fish, specified in Section 2.1 (i.e., "*their average wet weight must be between 0.3 and 2.5 g*"), was met. It must also be used to confirm that the requirement for fish-loading density in test solutions specified in Section 4.2 (i.e., "*the loading of fish in each test vessel must not exceed a density of 0.5 g/L*") was met.

All surviving fish (including the controls) used in the test must be disposed of in a humane manner at the end of the test. Overdosing the fish with an anaesthetic such as tricaine methanesulphonate is recommended.

Procedure for a Single-concentration Test to Determine the Mortality Rate at 96 Hours

All conditions, procedures, and facilities specified in Sections 1, 2, 3, 4, 7, and 8 apply to the procedure for testing a single concentration of effluent.

This procedure uses one concentration of effluent, 100% unless otherwise specified, plus a control. The use of replicate solutions (e.g., three replicates of the 100% concentration and three replicate control solutions, using 10 fish in each replicate solution) is recommended for this test, to provide greater confidence in the test results and their interpretation. At least ten fish must be exposed to each solution of effluent and control. The test is invalidated if >10% of the control fish (combined data, if replicates are used) exhibit atypical/stressed behaviour and/or mortality.

The endpoint for this test is percentage mortality at 96 h. Mortality of 50% is commonly used. For example, the 1992 Federal Pulp and Paper Effluent Regulations promulgated under the Federal Fisheries Act (Government of Canada, 1992) define an effluent as failing this test if the effluent at 100% concentration kills more than 50% of the fish.

Procedure for a Multi-concentration Test to Determine the 96-h LC50

All conditions, procedures, and facilities specified in Sections 1, 2, 3, 4, 7, and 8 apply to this procedure.

At least five concentrations of effluent plus a control (dilution water only) must be used in tests to estimate an LC50. At least ten fish must be exposed to each test concentration, including the undiluted (100%) concentration and the control. The highest concentration must be full-strength (100%) effluent, and each successive concentration must have at least 50% of the strength of the next higher one. A geometric (logarithmic) series is beneficial (e.g., percent concentrations such as 100, 50, 25, 12.5, 6.3). Concentrations may be based on other proportions or on standard dilution series (see EC, 1990a; Appendix D).

Since this LC50 test must include full-strength (100%) effluent as the highest concentration, the single-concentration endpoint of percent mortality in 100% effluent at 96 h (see Section 5) can also be determined from the results of this test.

Replicates of each concentration may be used but are not required. If replicates are used, their data are combined for calculating the LC50. The precision of the estimate of LC50 increases with the number of fish exposed to each test concentration, although the accuracy of this estimate is not necessarily improved.

The 96-h LC50 and its 95 % confidence limits should be calculated and the method of calculation reported. Computer programs for calculating LC50 and confidence limits are available (EC, 1990a) and should be used. A recommended program is available from Environment Canada (address in Appendix) for copying onto a user-supplied diskette, through the courtesy of C.E. Stephan (Stephan, 1977). A check of the computer-derived LC50 should be made by examining a plot on logarithmic-probability scales, of percent mortalities at 96 h for the various effluent concentrations (see EC, 1990a).

The test is invalidated if >10% of the control fish (combined data if replicates are used) exhibit atypical/stressed behaviour and/or mortality.

Procedure for Tests with a Reference Toxicant

A reference toxicant must be used to assess the relative sensitivity of the population of fish used in the toxicity test, and the precision and reliability of data produced by the laboratory personnel for that reference toxicant under standardized test conditions. The selected reference chemical(s) must be tested at least once during each calendar month when an effluent is tested, and upon acclimation of a new batch of fish. Fish used in the reference toxicity test conducted in conjunction with a test for determining the acute lethality of effluent must be from the same group held at the laboratory and used in the effluent test. The procedures and conditions to be followed are identical to those in Section 4 and as described in Environment Canada (1990a,c), except that a reference chemical is measured out and tested, instead of an effluent. The control/dilution water used routinely in effluent tests must also be used for the reference test.

Reagent-grade phenol and/or zinc sulphate ($ZnSO_4 \cdot 7H_2O$) are recommended for use as reference toxicants. The 96-h LC50 should be determined for the reference toxicant(s) used and expressed as mg/L based on phenol or zinc (Zn^{++}) (see EC, 1990a). Stock solutions of phenol must be made up on the day of use, and those for zinc should be prepared fresh on the day of use or stored in the dark at pH 3 to 4.

Concentrations of reference toxicant in all stock solutions should be measured chemically using appropriate methods (APHA *et al.*, 1998). Upon preparation of the test solutions, aliquots are to be taken from at least the control, low, middle, and high concentrations, and analyzed directly or stored for future analysis should the LC50 be atypical (i.e., outside warning limits). If stored, sample aliquots must be held in the dark at 4 ± 2°C. Both zinc and phenol aliquots should be preserved (APHA *et al.*, 1998) before storage. Stored aliquots requiring chemical measurement should be analyzed promptly upon completion of the toxicity test. It is desirable, but not required, to measure concentrations in the same solutions at the end of the test after completing biological observations. Calculations of LC50 should be based on measured concentrations if they are appreciably (i.e., ≥20%) different from nominal ones.

Once sufficient data are available (EC, 1990c), a warning chart which plots values for LC50 must be prepared, and continually updated, for each reference toxicant used. The warning chart should plot logarithm of concentration on the vertical axis against date of the test or test number on the horizontal axis. Each new LC50 for the reference toxicant should be compared with the established warning limits of the chart; the LC50 is acceptable if it falls within the warning limits. All calculations of mean and standard deviation must be made on the basis of log(LC50). The mean of log(LC50), together with its upper and lower warning limits (± 2 SD) as calculated by using the available values of log(LC50), are recalculated with each successive LC50 until the statistics stabilize (USEPA, 1985a; EC, 1990c).

The warning chart may be constructed by simply plotting mean and ± 2 SD as the logarithms, or if desired, by converting them to arithmetic values and plotting LC50 and ± 2 SD on a logarithmic scale of concentration. If a particular LC50 fell outside the warning limits, the sensitivity of the test fish and the performance and precision of the test would be suspect. A check of all acclimation and test conditions is required under these circumstances. Depending on the findings, further acclimation and re-evaluation of the fish with one or more reference toxicants should be undertaken, or a new population of fish should be procured and acclimated for use in subsequent toxicity tests with effluent and reference toxicant(s).

Reporting Requirements

The following is a summary of reporting and record-keeping requirements associated with this *reference method.* Further details or explanation can be found within previous sections of this method.

Unless otherwise specified by Environment Canada, all items listed in Section 8.1 must be reported to Environment Canada for each toxicity test that is initiated. The information is to be provided in accordance with pertinent regulations, and in a manner and format specified by Environment Canada* (i.e., manual or electronic, transmission mode, form and content).

Information additional to that in Section 8.1, such as that required by or distinctive to a regulation, or information that is necessary to clarify reporting and data assessment, may also be specified by Environment Canada.

Unless otherwise specified by Environment Canada, those items listed under Section 8.2 must be recorded and held on file for a period of five years. This information is to be provided as and when requested by Environment Canada. It will be required on a less frequent basis, such as during an audit or investigation.

8.1 Data to be Reported

8.1.1 Effluent

- name and location of operation generating the effluent;
- date and time of sampling;
- type of sample (e.g., "whole effluent from plant", "final mill effluent", "discharge from emergency spill lagoon", "leachate");
- brief description of sampling point;
- sampling method (e.g., "grab", "batch", "24-h composite with sub-samples at 1-h intervals"); and
- person collecting the sample.

8.1.2 Test Facilities and Conditions

- test type and method; e.g., "single-concentration test method as specified in the second (December 2000) edition of EPS 1/RM/13";
- indication of any deviation from one or more "must" requirements delineated in Sections 2 to 7 of this report, including details as to the deviation;
- name and city of testing laboratory;
- percent mortality of fish in stock tank(s) from which test fish are taken, as recorded daily (or, as a minimum, for five of the seven days spanning a weekly period) for the seven-day period immediately preceding the test;
- species of test organism;
- date and time for start of definitive test;

* Contact an office listed in the Appendix for details.

- person(s) performing the test and verifying the results;
- the pH, temperature, dissolved oxygen, and conductivity of unadjusted, undiluted effluent, just before preparing test solutions;
- confirmation that no adjustment of sample or solution pH occurred; indication of procedure used for any pH adjustment if both pH-adjusted and non-adjusted tests were run (see Section 4.2);
- indication of aeration of test solutions (rate, time) before introduction of fish; rate of aeration throughout the test;
- concentrations and volumes tested, including controls, and indication of any replication;
- measurements of dissolved oxygen, pH, and temperature determined for each test solution including control(s) at the beginning and end of the test, as well as conductivity of each test solution at the beginning of the test;
- number of fish added to each test vessel;
- mean fork length of control fish at the end of the test, together with the range of the values measured;
- mean wet weight of individual control fish at the end of the test; and
- estimated loading density (g/L) of fish in test solutions.

8.1.3 Results

- number of mortalities of fish in each test solution including the control(s), at 96 h; number of control fish showing atypical/ stressed behaviour;
- mean mortality rate in solutions of effluent and control water, if a single-concentration or multi-concentration test is performed using replicate solutions; mean number of control fish showing atypical/stressed behaviour if replicate control solutions;
- estimate of 96-h LC50 and 95% confidence limits in multi-concentration tests, if statistically achievable; indication of statistical method (e.g., log-probit, moving average) on which result is based; and
- most recent 96-h LC50 (with 95% confidence limits) for reference toxicity test(s) performed with the group of fish used in the effluent test; reference chemical(s); date test initiated; historic geometric mean LC50 and warning limits (± 2 SD).

*8.2 Data to be Held on File**

8.2.1 Effluent

- all information (e.g., code, sample description, date/time of sampling) affixed to label(s) on sample container(s);
- volume of sample;
- transport and storage conditions (times; temperature on receipt at laboratory and during storage; indication if sample frozen or partially frozen on arrival);

---

* To be stored for a five-year period at the testing facility and/or the offices of the discharger. Some of this information may be common to a series of tests, and recorded and held on file as a general report.

- appearance and other properties (observations on colour, turbidity, odour, floating or settleable material);

- colour change, precipitation, flocculation, release of volatiles or other changes when making up test solution(s); and

- procedures and results for any chemical analyses performed on the effluent, if available (e.g., suspended solids content, hardness).

8.2.2 Test Facilities and Conditions

- address of testing laboratory;

- description of rearing/acclimation and test facilities including general layout of each and means of isolation;

- normal holding and acclimating conditions (containers, location, lighting, temperatures including maximum rate of change, aeration, volumes and flows of water, procedure for water renewal, numbers and densities of fish, handling procedures, food type, ration and frequency of feeding, disease incidence and treatment if any, weekly cumulative percent mortality);

- source of test fish;

- brief history of test-specific conditions and procedures for holding and acclimating fish (e.g., times, water source, and characteristics such as temperature, pH and dissolved oxygen content, food type and ration, disease incidence and treatment, weekly cumulative percent mortality) if different from usual practice;

- description of source(s) of water used for rearing and acclimating fish and as control/dilution water;

- pretreatment of acclimation and control/dilution water, if any (e.g., adjustment of temperature, aeration rate and duration, quantity of any chemical added);

- quality (mean and range values) of acclimation and control/dilution water as measured for source water and within holding tank(s); to include hardness, pH, conductivity, dissolved oxygen content, and total residual chlorine (if dechlorinated municipal drinking water); preferably also total dissolved gases, alkalinity, solids, organic carbon, colour, mineral ions, metals, ammonia, nitrite, and organophosphorus pesticides;

- systems to regulate light and temperature;

- light source, photoperiod, and past measures of intensity at rearing/acclimation tanks and at surface of test vessels;

- description of test vessels (size, shape, and material), covers and baskets (if used for inspecting fish), and routine cleaning procedures for each;

- procedures used to randomize the introduction of fish to test vessels;

- procedure and apparatus for aeration of test solutions;

- all measurements of fork length and wet weight of individual fish used in test (together with the mean value and range of values for each sample);

- depth of test solutions; appearance of solutions, including any changes evident during test;

- test concentrations of reference toxicant(s), both nominal and measured; indication of data set used to estimate LC50; and

- any measurements of water quality in test solutions not included in data reported (Section 8.1.2).

8.2.3 Results

- any observations of mortalities of fish not included in data reported (e.g., at 24, 48, and 72 h) (see Section 8.1.3);

- observations of fish behaviour and appearance recorded for each test solution during the test; and

- any manual plot(s) of data used to verify a computer-derived LC50.

References

APHA, AWWA, and WEF (American Public Health Association, American Water Works Association, and Water Environment Federation), *Standard Methods for the Examination of Water and Wastewater,* 20th ed., L. Clesceri, A. Greenber, and A. Eaton (eds.), APHA, AWWA, and WEF, Washington, DC (1998).

ASTM (American Society for Testing and Materials), "Standard Practice for Conducting Acute Toxicity Test with Fishes, Macroinvertebrates, and Amphibians", ASTM, Philadelphia, PA, Report E729-80, 25 p. (1980).

BHSC (British Health and Safety Commission), "Methods for the Determination of Ecotoxicity", Approved Code of Practice, Notification of New Substances Regulations 1982, BHSC, London, England, Report COP 8, 13 p. (1982).

Craig, G.R., K. Flood, J. Lee, and M. Thomson, "Protocol to Determine the Acute Lethality of Liquid Effluents to Fish", Ontario Ministry of Environment, Rexdale, Ontario, 9 p. (1983).

EC (Environment Canada), "Biological Test Method: Acute Lethality Test Using Rainbow Trout", Conservation and Protection, Ottawa, Ontario, Report EPS 1/RM/9 including May 1996 Amendments, 51 p. (1990a).

EC (Environment Canada), "Biological Test Method: Reference Method for Determining Acute Lethality of Effluents to Rainbow Trout", Conservation and Protection, Ottawa, Ontario, Report EPS 1/RM/13 including May 1996 amendments, 18 p. (1990b).

EC, "Guidance Document for Control of Toxicity Test Precision Using Reference Toxicants", Conservation and Protection, Ottawa, Ontario, Report EPS 1/RM/12 (1990c).

EPS (Environmental Protection Service), "Pulp and Paper Effluent Regulations", Environment Canada, Regulations, Codes, Protocols, Ottawa, Ontario, Report EPS 1-WP-72-1 (1971).

EPS, "Guidelines on the Use and Acceptability of Oil Spill Dispersants", Environment Canada, Ottawa, Ontario, Report EPS 1-EE-73-1 (1973).

EPS, "Petroleum Refinery Effluent Regulations and Guidelines", Environment Canada, Regulations, Codes, and Protocols, Ottawa, Ontario, Report EPS 1-WP-74-1 (1974).

EPS, "Metal Mining Liquid Effluent Regulations and Guidelines", Environment Canada, Regulations, Codes, and Protocols, Ottawa, Ontario, Report EPS 1-WP-77-1 (1977a).

EPS, "Meat and Poultry Products Plant Liquid Effluent Regulations and Guidelines", Environment Canada, Regulations, Codes, and Protocols, Ottawa, Ontario Report EPS 1-WP-77-2 (1977b).

EPS, "Potato Processing Plant Liquid Effluent Regulations and Guidelines", Environment Canada, Regulations, Codes, and Protocols, Ottawa, Ontario, Report EPS 1-WP-77-4 (1977c).

EPS, "Standard Procedure for Testing the Acute Lethality of Liquid Effluents", Environment Canada, Regulations, Codes, and Protocols, Ottawa, Ontario, Report EPS 1-WP-80-1 (1980).

EPS, "Guidelines on the Use and Acceptability of Oil Spill Dispersants", Environment Canada, Regulations, Codes, and Protocols, Ottawa, Ontario, Report EPS 1-EP-84-1 (1984).

Government of Canada, "Fisheries Act: Pulp and Paper Effluent Regulations", *Canada Gazette,* Part II, Vol. 126, No. 11, p. 1967–2006 (1992).

Leitritz, E. and R.C. Lewis, "Trout and Salmon Culture (Hatchery Methods)", California Fish Bulletin, No. 164, University of California, Division of Agricultural Sciences, Berkeley, CA, 197 p. (1976).

McGuinness, E.J., "Procedures Manual, Aquatic Bioassay Service Projects", Report AECV-82M1, Alberta Environmental Centre, Vegreville, Alberta, 56 p. (1982).

OECD (Organization for Economic Cooperation and Development), "Guideline for Testing of Chemicals – Fish, Acute Toxicity Test", OECD, Paris, France, Document No. 203, 12 p. (1984).

OME (Ontario Ministry of Environment), "Protocol for Single Concentration Acute Lethality Tests on Liquid Effluent Samples using Rainbow Trout", OME, Rexdale, Ontario, 10 p. (1989).

Rocchini, R.J., M.J.R. Clark, A.J. Jordan, S. Horvath, D.J. McLeay, J.A. Servizi, A. Sholund, H.J. Singleton, R.G. Watts, and R.H. Young, "Provincial Guidelines and Laboratory Procedures for Measuring Acute Lethal Toxicity of Liquid Effluents to Fish", BC Ministry of Environment, Victoria, BC, 18 p. (1982).

Stephan, C.E., "Methods for Calculating an LC50", p. 65–84, in: *Aquatic Toxicology and Hazard Evaluation,* F.L. Mayer and J.L. Hamelink (eds.), ASTM STP 634, American Society for Testing and Materials, Philadelphia, PA (1977).

UKWRC (United Kingdom Water Research Centre), "Acute Toxicity to Fish. Determination of the 96-h LC50 of Test Substances to the Rainbow Trout Under Flow-through Conditions", Water Research Centre, Great Britain, Medmenham, UK, Standard Operating Procedure, No. EP 120 02, 19 p. (1983).

USEPA (United States Environmental Protection Agency), "Methods for Measuring the Acute Toxicity of Effluents to Freshwater and Marine Organisms", W.H. Peltier and C.I. Weber (eds.), USEPA, Cincinnati, Ohio, Report EPA/600/4-85/013, 216 p. (1985a).

USEPA, "Acute Toxicity Test for Freshwater Fish. Standard Evaluation Procedure", USEPA, Hazard Evaluation Division, Washington, DC, Report EPA-540/9-85-006, 12 p. (1985b).

USEPA (United States Environmental Protection Agency), "Methods for Aquatic Toxicity Identification Evaluations, Phase I. Toxicity Characterization Procedures", Second Edition, Report EPA/600/6-91/003, USEPA, Duluth, MN (1991).

Members of the Inter-Governmental Environmental Toxicity Group and Addresses of Environment Canada's Headquarters and Regional Offices

Members of the Inter-Governmental Environmental Toxicity Group (as of December, 2000):

Federal (Environment Canada)

C. Blaise
Centre St. Laurent
Montreal, Québec

M. Bombardier
Environmental Technology Centre
Ottawa, Ontario

U. Borgmann
National Water Research Institute
Burlington, Ontario

J. Bruno
Pacific Environmental Science Centre
North Vancouver, British Columbia

C. Buday
Pacific Environmental Science Centre
North Vancouver, British Columbia

T. Corbin
Atlantic Environmental Science Centre
Moncton, New Brunswick

K. Doe
Atlantic Environmental Science Centre
Moncton, New Brunswick

G. Elliott
Environmental Protection Service
Edmonton, Alberta

M. Fennell
Pacific Environmental Science Centre
North Vancouver, British Columbia

F. Gagné
Centre St. Laurent
Montreal, Québec

M. Harwood
Centre St. Laurent
Montreal, Québec

P. Jackman
Atlantic Environmental Science Centre
Moncton, New Brunswick

N. Kruper
Environmental Protection Service
Edmonton, Alberta

D. MacGregor
Environmental Technology Centre
Ottawa, Ontario

D. Moul
Pacific Environmental Science Centre
North Vancouver, British Columbia

W.R. Parker
Environmental Protection Service
Fredericton, New Brunswick

L. Porebski
Marine Environment Division
Hull, Québec

D. Rodrique
Environmental Technol. Advancement Directorate
Hull, Québec

R. Scroggins
Environmental Technology Centre
Ottawa, Ontario

A. Steenkamer
Environmental Technology Centre
Ottawa, Ontario

G. van Aggelen (Chairperson)
Pacific Environmental Science Centre
North Vancouver, British Columbia

R. Watts
Pacific Environmental Science Centre
North Vancouver, British Columbia

P. Wells
Environmental Conservation Service
Dartmouth, Nova Scotia

Federal (Fisheries & Oceans Canada)

R. Roy
Insitute Maurice-Lamontagne
Mont-Joli, Québec

Provincial

C. Bastien
Ministère de l'Environnement du Québec
Ste. Foy, Québec

B. Bayer
Manitoba Environment
Winnipeg, Manitoba

D. Bedard
Ontario Ministry of Environment
Rexdale, Ontario

M. Mueller
Ontario Ministry of Environment
Rexdale, Ontario

D. Poirier
Ontario Ministry of Environment
Rexdale, Ontario

J. Schroeder
Ontario Ministry of Environment
Rexdale, Ontario

Addresses for Environment Canada's Regional and Headquarters' Offices

Headquarters
351 St. Joseph Blvd., Place Vincent Massey
Hull, Quebec
K1A 0H3

Atlantic Region
15th Floor, Queen Square
45 Alderney Drive
Dartmouth, Nova Scotia
B2Y 2N6

Quebec Region
8th Floor
105 McGill Street
Montreal, Québec
H2Y 2E7

Ontario Region
4905 Dufferin Street, 2nd Floor
Downsview, Ontario
M3H 5T4

Prairie and Northern Region
Twin Atria No. 2, Room #210
4999 - 98 Avenue
Edmonton, Alberta
T6B 2X3

Pacific and Yukon Region*
224 West Esplanade Street
North Vancouver, British Columbia
V7M 3H7

---

* A computer program for calculating LC50s is available for copying onto an IBM-compatible diskette supplied by the user, by contacting the Environmental Toxicology Section at this address.

Méthode d'essai biologique : méthode de référence pour la détermination de la létalité aiguë d'effluents chez la truite arc-en-ciel

Section de l'élaboration des méthodes et des applications
Centre de technologie environnementale
Environnement Canada
Ottawa (Ontario)

Méthode de référence SPE 1/RM/13
DEUXIÈME ÉDITION
Décembre 2000

Données de catalogage avant publication de la Bibliothèque nationale du Canada

Vedette principale au titre :

Méthode d'essai biologique : Méthode de référence
pour la détermination de la létalité aiguë d'effluents
chez la truite arc-en-ciel

Deuxième éd.
(Méthode de référence; SPE 1/RM/13)
Texte en anglais et en français disposé tête-bêche.
Titre de la p. de t. addit.: Biological test method :
Reference method for determining acute lethality
of effluents to rainbow trout.
Comprend des références bibliograhiques.
ISBN 0-660-61615-7
No. de cat. En49-24/1-13-2001

1. Truite arc-en-ciel -- Mortalité -- Essais -- Normes -- Canada.
2. Truite arc-en-ciel -- Microbiologie -- Normes -- Canada.
3. Qualité effluents -- Essais -- Normes -- Canada.
4. Toxicologie -- Méthodologie.
I. Centre de technologie environnementale (Canada), Section de
l'élaboration des méthodes et des applications.
II. Canada. Environnement Canada.
III. Coll.: Rapport (Canada. Environnement
Canada); SPE 1/RM/13.

QH90.57B56 2001 597.5'7 C2001-980086-XF

No. de cat. En49-24/1-13-2001
ISBN 0-660-61615-7

Commentaires des lecteurs

Pour formuler des commentaires sur le contenu de la présente méthode de référence, s'adresser à:

Richard Scroggins
Section de l'élaboration des méthodes et des applications
Centre de technologie environnementale
Environnement Canada
3439, Chemin River sud
Ottawa (Ontario)
K1A 0H3

Avis de révision

Le présent document a été révisé par le personnel de la Direction générale de l'avancement des technologies environnementales d'Environnement Canada et a été approuvé pour publication. La mention de marques de commerce ou de produits commerciaux ne constitue nullement une recommandation de la part d'Environnement Canada; d'autre produits de valeur semblable sont disponibles.

Résumé

Ce rapport fournit des méthodes normalisées ou de référence explicites pour mesurer la létalité d'effluents pour la truite arc-en-ciel (Oncorhynchus mykiss). *Des instructions précises sont données pour exécuter des essais de toxicité létale avec des échantillons d'effluent. Les lignes directrices données dans la méthode générale "Essai de létalité aiguë chez la truite arc-en-ciel" (EC, 1990a, avec modifications apportées en mai 1996) ont été incluses.*

Le présent rapport est la deuxième édition de la Méthode de référence SPE 1/RM/13 publiée en juillet 1990, à laquelle des modifications ont été apportées en mai 1996 (EC, 1990b - comprend les modifications de mai 1996). Cette méthode remplace la version précédente et doit être appliquée comme la méthode de référence actuelle d'Environnement Canada pour la détermination de la létalité aiguë d'effluents chez la truite arc-en-ciel.

Trois méthodes sont exposées: 1) pour un essai à concentration unique portant sur l'effluent non dilué, sauf indication contraire; 2) pour un essai à concentrations multiples visant à déterminer la concentration létale 50 (CL50) de l'effluent; 3) pour un essai portant sur un produit toxique de référence. Le lecteur trouvera des instructions pour la détention des truites au laboratoire, les installations et l'approvisionnement en eau, la manipulation et le stockage des échantillons, la préparation des solutions, les conditions de l'essai, les observations à faire, les résultats de l'essai et les méthodes de calcul connexes ainsi que l'utilisation de produits toxiques de référence.

Avant-propos

Ce rapport fait partie de la collection des ***méthodes de référence*** *pour la mesure et l'évaluation de l'effet ou des effets toxiques sur une seule espèce d'organismes aquatiques ou terrestres exposés à des échantillons de matières ou de substances d'essai dans des conditions de laboratoire définies et contrôlées.*

Dans le présent document, une ***méthode de référence*** *est définie comme une méthode d'essai biologique spécifique pour conduire un essai de toxicité, c'est-à-dire une méthode d'essai qui donne un ensemble d'instructions et de conditions détaillées dans un document écrit. Contrairement à d'autres méthodes d'essai biologique polyvalentes (générales) publiées par Environnement Canada, l'utilisation d'une méthode de référence se résume fréquemment aux exigences associées à des règlements précis.*

Les ***méthodes de référence*** *sont celles qui ont été développées et publiées par Environnement Canada (EC) et sont privilégiées :*

- *en vue de l'application des règlements dans les laboratoires de toxicité environnementale des agences fédérales et provinciales;*
- *pour des essais de toxicité à des fins de réglementation confiés par Environnement Canada à des laboratoires externes ou demandés par des agences externes ou l'industrie;*
- *pour incorporation dans les règlements ou permis fédéraux, provinciaux ou municipaux, comme exigence de surveillance réglementaire;*
- *comme base pour la stipulation d'instructions très explicites.*

Table des matières

Résumé v
Avant-propos vi
Glossaire ix
Remerciements xii

Section 1
Introduction 1

Section 2
Organismes et détention 3
2.1 Espèce et source 3
2.2 Détention et acclimatation 3
2.3 Eau 4
2.4 Conditions physico-chimiques 5

Section 3
Installations 6

Section 4
Mode opératoire général pour la détermination de la létalité aiguë d'effluents 7
4.1 Étiquetage, transport et stockage des échantillons 7
4.2 Conditions de l'essai 8
4.3 Préparation des solutions d'essai 8
4.4 Mise en route de l'essai 9
4.5 Observations et mesures 10

Section 5
Mode opératoire pour la détermination du taux de mortalité après 96 h dans un essai à concentration unique 11

Section 6
Mode opératoire pour la détermination de la CL50–96 h dans un essai à concentrations multiples 12

Section 7
Mode opératoire pour les essais avec un produit toxique de référence 13

Section 8
Présentation des résultats 15
8.1 Données à consigner au procès-verbal 15
8.1.1 Effluent 15
8.1.2 Installations et conditions de l'essai 15
8.1.3 Résultats 16
8.2 Données à verser au dossier 17
8.2.1 Effluent 17
8.2.2 Installations et conditions de l'essai 17
8.2.3 Résultats 18

Références 19

Annexe
Membres du Groupe intergouvernemental de la toxicité environnementale et adresses de l'administration centrale et des bureaux régionaux d'Environnement Canada 21

Glossaire

Toutes les définitions ci-après s'inscrivent dans le contexte des modes opératoires énoncés dans le présent rapport; les autres définitions figurant dans le document d'accompagnement détaillé (EC, 1990a, avec modifications apportées en mai 1996) s'y appliquent aussi.

Termes grammaticaux

L'auxiliaire *doit (doivent)* et *il faut* sont utilisés pour exprimer une obligation absolue.

L'auxiliaire *devrait (devraient)* et le mode conditionnel *(il faudrait*, etc.) sont utilisés pour indiquer que la condition ou la méthode en cause est recommandée et doit être respectée dans la mesure du possible.

L'auxiliaire *peut (peuvent)* indique qu'on est autorisé à faire une chose ou en mesure de le faire.

Termes techniques

Acclimatation - Adaptation physiologique à un niveau précis d'une ou de plusieurs variables environnementales, par exemple la température. Ce terme s'applique généralement à des conditions contrôlées en laboratoire.

Aigu - Survenant dans un bref délai, normalement ≤96 h dans le cas des poissons.

Alevin - Poisson éclos récemment, qui n'a pas commencé à se nourrir seul et dont le sac vitellin est évident (à des fins nutritives). L'alevin est souvent appelé «larve vésiculée».

Alevin nageant - Jeune poisson ayant dépassé le stade larvaire et ayant commencé à se nourrir seul activement.

CL50 (concentration létale médiane) - Concentration dans l'eau d'une substance (dans le cas qui nous occupe, un effluent) que l'on a estimée être létale pour 50 % des organismes soumis à l'essai. La CL50 et ses limites de confiance à 95 % sont généralement obtenues par l'analyse statistique des pourcentages de mortalité dans plusieurs concentrations d'essai, après une période fixe d'exposition qui doit être précisée (p. ex., CL50–96 h).

Conductivité - Expression numérique de la capacité d'une solution aqueuse de transporter un courant électrique. Cette capacité dépend des concentrations des ions en solution, de leur valence et de leur mobilité, ainsi que de la température de la solution. La conductivité s'exprime normalement en millisiemens par mètre, une unité SI (système international), ou en μmho/cm (1 mS/m = 10 μmhos/cm).

Contrôle - Traitement reproduisant l'ensemble des conditions et facteurs qui pourraient influencer les résultats d'une enquête ou d'une étude, à l'exception de la condition particulière faisant l'objet de cette étude. Dans les essais de toxicité aquatique, le contrôle doit reproduire toutes les conditions du ou des traitements d'exposition, mais ne pas renfermer la substance à expérimenter. Le contrôle est utilisé pour établir l'absence de toxicité mesurable en raison de conditions de base de l'essai, telles que la température, la qualité de l'eau de dilution, l'état de santé des organismes d'essai ou les effets dus à leur manipulation.

Dureté - Concentration des cations dans l'eau qui réagira avec un savon sodique pour précipiter un résidu insoluble. En général, la dureté est une mesure des concentrations d'ions calcium et magnésium dans l'eau, exprimée en mg/L de carbonate de calcium.

Eau de contrôle/de dilution - Eau utilisée pour diluer l'échantillon d'effluent et pour l'essai de contrôle.

Eau de dilution - Eau utilisée pour diluer la substance à expérimenter afin d'en préparer différentes concentrations pour l'essai de toxicité.

Eau déchlorée - Eau chlorée (généralement, eau potable du réseau d'alimentation municipal) qui a été traitée afin d'en éliminer le chlore et les composé chlorés.

Effluent - Tout déchet liquide (industriel ou urbain) rejeté dans l'environnement aquatique.

Essai à renouvellement continu - Essai de toxicité pendant lequel les solutions des réservoirs d'essai sont renouvelées en continu par l'apport constant d'une solution fraîche ou par un apport intermittent fréquent.

Essai à renouvellement périodique - Essai de toxicité pendant lequel les solutions d'essai sont renouvellées périodiquement, généralement aux 24 heures. *Syn. renouvellement, semi-statique, à renouvellement séquentiel.*

Essai statique - Essai de toxicité pendant lequel les solutions d'essai ne sont pas renouvelées.

Évident - Clairement discernable dans les conditions d'essai utilisées.

Fingerling - Jeune poisson de moins d'un an se nourissant seul.

Létal - Qui entraîne la mort par action directe. Dans le cas d'un poisson, on entend par «mort» la cessation de tous les signes visibles de mouvement ou d'activité.

Longueur à la fourche - Longueur du poisson, du museau (gueule fermée) jusqu'à la fourche de la queue.

Lux - Unité SI d'éclairement, mesurant l'intensité lumineuse par mètre carré. Un lux = 0,0929 pied-bougie; un pied-bougie = 10,76 lux.

Méthode de référence - Méthode d'essai biologique spécifique pour réaliser un essai de toxicité, c'est-à-dire une méthode d'essai comportant un ensemble d'instructions et de conditions explicites précisément décrites dans un document écrit. Contrairement à d'autres méthodes d'essais biologiques polyvalentes publiées par Environnement Canada, une *méthode de référence* est fréquemment utilisée aux seules fins de tester les exigences associées à des règlements précis; ou tester dans le but de déterminer s'il y a violation des dispositions générales de la Loi canadienne sur les pêches.

pH - Logarithme négatif de l'activité des ions hydrogène, mesurée par leur concentration en équivalents grammes par litre. Cette valeur exprime le degré ou l'intensité des réactions acides et alcalines selon une échelle de 0 à 14, où le nombre 7 représente la neutralité et les nombres inférieurs correspondent, en ordre décroissant, à des réactions acides de plus en plus fortes. Les chiffres supérieurs à 7 indiquent, en ordre croissant, des réactions basiques ou alcalines de plus en plus fortes.

Photopériode - Durée d'éclairement et d'obscurité au cours d'un cycle de 24 h.

Prétraitement - Dans le présent rapport, traitement d'un échantillon ou de sa dilution avant l'exposition des poissons.

Produit toxique de référence - Étalon chimique utilisé pour évaluer la sensibilité des organismes soumis à l'essai afin d'établir le degré de confiance dans les données de toxicité obtenues pour une matière ou un produit soumis à un essai. Dans la plupart des cas, un essai de toxicité avec un produit de référence est exécuté pour évaluer la sensibilité des organismes au moment où l'on évalue la matière d'essai et la précision des résultats obtenus par le laboratoire avec ce produit chimique.

Salinité - Quantité totale, en grammes, de substances solides dissoutes dans 1 kg d'eau. Elle se détermine après conversion de tous les carbonates en oxydes, après remplacement de tous les bromures et iodures par des chlorures et après oxydation de toutes les matières organiques. La salinité peut aussi se mesurer directement grâce à un salinimètre ou conductimètre ou par d'autres moyens (voir APHA *et al.,* 1998). Elle est habituellement rapportée en grammes par kilogramme ou en parties par mille (‰).

Sublétal - Nocif pour les poissons, mais en deçà du niveau qui entraîne directement la mort pendant la durée de l'essai.

Toxicité - Potentiel ou capacité propre d'une substance de provoquer un ou des effets nocifs chez les poissons ou autres organismes vivants. Le (ou les) effet(s) peu(ven)t être létal(aux) ou sublétal(aux).

Remerciements

La première édition de ce rapport, imprimée en juillet 1990 et à laquelle des modifications ont été apportées en mai 1996 (EC, 1990b), a été réalisée par D.J. McLeay (McLeay Associates Ltd., West Vancouver, C.-B.) et J.B. Sprague (Sprague Associates Ltd., Guelph, Ontario). M. R. Scroggins et M. G. Sergy (SPE, Environnement Canada) on fait fonction de responsables scientifiques officiels et ont pourvu aux orientations et à l'aide technique. La première édition a été établie en fonction des objectifs, des besoins et des commentaires d'un sous-comité du Service de la protection de l'environnement (SPE) formé de: N. Bermingham (SPE, Montréal); C. Blaise (SPE, Longueuil); G. Coulombe (SPE, Montréal); K. Doe (SPE, Dartmouth); G. Elliott (SPE, Edmonton); D. MacGregor (SPE, Ottawa); J. MacLatchy (SPE, Ottawa); W. Parker (SPE, Dartmouth); D. Robinson (Pêches et Océans Canada, Ottawa); T. Ruthman (SPE, Ottawa); R. Scroggins (SPE, Ottawa); G. Sergy (SPE, Edmonton) et R. Watts (SPE, North Vancouver). Les représentants des provinces au sein du Groupe intergouvernemental de la toxicité aquatique, en place en 1990, ont apporté à l'époque une contribution et des commentaires fort utiles
(G. Joubert, ministère de l'Environnement du Québec; G. Westlake, ministère de l'Environnement de l'Ontario; W. Young, ministère de l'Environnement et de la Sécurité publique du Manitoba;
K. Lauten, ministère de l'Environnement et de la Sécurité publique de la Saskatchewan;
J. Somers, ministère de l'Environnement de l'Alberta; S. Horvath and G. van Aggelen, ministère de l'Environnement de la Colombie-Britannique). Les personnes suivantes ont également revu les versions préliminaires de la première édition et contribué à sa publication : D. Vaughan (SPE, Dartmouth); S. Wade (SPE, Dartmouth); A. Beckett (SPE, Edmonton); F. Zaal (SPE, Edmonton); S. Yee (SPE, North Vancouver); D. Moul (SPE, North Vancouver); V. Zitko (Pêches et Océans Canada, St. Andrews); V. Cairns (Pêches et Océans Canada, Burlington); S. Samis (Pêches et Océans Canada, Vancouver); R. Martin (ministère de l'Environnement et des Terres de Terre-Neuve, St. John's); J. Lee (ministère de l'Environnement de l'Ontario, Rexdale); J. Moores (ministère de l'Environnement de l'Alberta, Vegreville); P. Martel (Institut canadien de recherches sur les pâtes et papiers, Pointe-Claire); Y. Bois (Technitrol · Eco Inc., Pointe-Claire); C. R. Cook (produits forestiers E.B. Eddy Ltée, Espanola); G. Craig (Beak Consultants Ltd., Brampton) ainsi que D. Bradley, D. Monteith, et J. Pickard (B.C. Research Corp., Vancouver). Nous tenons également à remercier l'équipe des laboratoires d'essai d'Environnement Canada (adresses en annexe).

Les personnes suivantes on contribué aux modifications apportées en mai 1996 à la première édition de cette méthode de référence (EC, 1990b), qui sont maintenant incorporées dans cette deuxième édition (avec des modifications si besoin était) : C. Bastien (ministère de l'Environnement du Québec, Sainte-Foy); B. Bayer (ministère de l'Environnement du Manitoba, Winnipeg); K. Doe (Service de la conservation de l'environnement - SCE - Moncton); G. Elliott (SPE, Edmonton); P. Jackman (SCE, St. John's); J. Miller (SPE, Ottawa); G. van Aggelen (SCE, North Vancouver); G. Westlake (ministère de l'Environnement de l'Ontario, Rexdale); et R. Watts (SCE, North Vancouver).

Cette deuxième édition a été préparée par D.J. McLeay (McLeay Associates Ltd.,Victoria) qui a reçu des conseils et de l'aide de R.P. Scroggins (Section de l'élaboration des méthodes et des applications, Environnement Canada, Ottawa). Les membres du Groupe intergouvernemental de la toxicité environnementale (voir annexe) ont contribué aux modifications introduites dans cette édition et les ont appuyées. Nous tenons à remercier tout spécialement les personnes suivantes à qui nous devons certaines des modifications et ajouts : K. Doe (SCE, Moncton); G. Elliot (SPE, Edmonton) et J. Schroeder (ministère de l'Environnement de l'Ontario, Rexdale).

Introduction

Le présent document remplace la première édition de la *méthode de référence* SPE 1/RM/13 (EC, 1990b) publiée en juillet 1990 et à laquelle des modifications ont été apportées en mai 1996. Cette deuxième édition reprend les insructions d'analyse, les lignes directrices et le libellé du rapport SPE 1/RM/13 tel que modifié en mai 1996, à l'exception de modifications ou d'ajouts qui ont été jugés nécessaires ou qui constituaient des améliorations prudentes selon Environnement Canada.

Le présent document décrit les modes opératoires à employer pour les essais de létalité aiguë sur la truite arc-en-ciel, tels que précisés par les gouvernements canadiens impliqués dans la surveillance et le contrôle de la pollution des effluents industriels ou municipaux. La truite arc-en-ciel est utilisée depuis trois décennies au Canada pour les essais d'effluents prévus par une série de règlements et de lignes directrices (SPE, 1971, 1973, 1974, 1977a-c, 1984; Gouvernement du Canada, 1992). Cette *méthode de référence* devrait être utilisée conjointement avec le rapport d'Environnement Canada (EC, 1990a, incluant les modifications apportées en 1996) qui présente les raisons à l'appui et des détails additionnels.

Par bien des aspects, les modes opératoires exposés ci-après sont semblables à des méthodes établies par les provinces canadiennes (McGuinness, 1982; Rocchini *et al.*, 1982; Craig *et al.*, 1983; MEO, 1989), à des méthodes employées aux États-Unis (ASTM, 1980; USEPA, 1985a; 1985b; APHA *et al.*, 1998), et à des techniques internationales (BHSC, 1982; UKWRC, 1983; OCDE, 1984). Ces méthodes se sont révélées fort utiles à la rédaction de l'ensemble du présent document, et elles constituent des sources de renseignements complémentaires très valables. Toutefois, aux fins de la réglementation, ce sont les modes opératoires présentés dans la présente *méthode de référence* qu'on devrait appliquer.

La truite arc-en-ciel *(Oncorhynchus mykiss*, auparavant *Salmo gairdneri)* doit être utilisée comme organisme d'essai lorsqu'on applique cette *méthode de référence*. Cette espèce, qui provient de l'Ouest de l'Amérique du Nord, fréquente maintenant les eaux de toutes les provinces canadiennes et est très répandue dans le monde entier. Elle prospère dan les eaux froides et douces, fréquente les eaux côtières de l'Atlantique et du Pacifique et est couramment élevée dans des écloseries et des établissements commerciaux d'aquiculture. Elle est devenue, dans le monde entier, le poisson d'eau froide étalon pour les essais de toxicité en eau douce, et on dispose à son sujet d'une banque de données toxicologiques d'une ampleur appréciable.

Trois modes opératoires de base sont décrits ci-après. Le premier porte sur une concentration unique d'effluent (l'échantillon non dilué, sauf indication contraire) et sur un contrôle, et il conviendrait à des essais comportant une cote de réussite ou d'échec. Le deuxième permet d'établir la concentration létale 50 (CL50) (c'est-à-dire,

de déterminer le degré de toxicité de l'effluent au moyen de diverses concentrations, dont l'échantillon non dilué). Le troisième s'applique à un essai à concentrations multiples avec un produit toxique de référence, qui permet d'évaluer la sensibilité des poissons soumis à l'essai à un produit toxique étalon ainsi que la précision des données produites par le laboratoire pour ce produit chimique.

L'essais doit porter sur des effluents contenant de l'eau douce ou ayant une salinité inférieure ou égale à 10 ‰, définie par une conductivité inférieure ou égale à 1400 mS/m à une température de 15° C. Les effluents de salinité supérieure à 10 ‰ qui sont rejetés dans des eaux douces devraient aussi faire l'objet d'essais effectués avec des truites arc-en-ciel acclimatées à l'eau douce. Pour les effluents de salinité supérieure à 10 ‰ qui sont rejetés directement dans des eaux réceptrices estuariennes ou marines, on devrait utiliser une espèce autorisée par le laboratoire régional d'Environnement Canada (adresses à l'annexe) et acclimatée à une eau de mer de salinité semblable à celle de l'effluent.

Organismes et détention

2.1 Espèce et source

La truite arc-en-ciel (*Oncorhynchus mykiss*) doit être utilisée pour les essais. On peut se servir d'alevins ayant atteint le stade de l'alimentation active depuis au moins deux semaines ou de fingerlings. Le poids frais moyen des poissons doit être compris entre 0,3 et 2,5 g, et le plus gros poisson ne devrait pas être plus de deux fois plus long que le plus petit poisson soumis au même essai.

On peut se procurer les poissons aux stades de l'oeuf embryonné, d'alevin ou de fingerling. Chaque groupe de poissons devrait provenir d'une écloserie agréée, reconnue pour être exempte de maladies et dotée d'un programme continu de surveillance de la santé des organismes et de certification. Leur obtention et leur expédition devraient être approuvées par des représentants régionaux du comité fédéral-provincial de transplantations d'espèces de Pêches et Océans Canada, dans les provinces où ce comité contrôle les mouvements des stocks de poissons. Pour obtenir des conseils sur les sources de truites, on peut communiquer avec les bureaux régionaux d'Environnement Canada (adresses à l'annexe).

2.2 Détention et acclimatation

Les poissons devraient être élevés dans des réservoirs et d'autres installations fabriqués de matériaux non toxiques comme l'acier inoxydable, la porcelaine, le polyester armé de fibre de verre, l'acrylique, le polyéthylène ou le polypropylène. Les oeufs et les alevins peuvent être incubés dans des plateaux d'alevinage à débit vertical ou dans des bassins à écoulement d'eau (Leitritz et Lewis, 1976). Les alevins et les fingerlings peuvent être élevés et acclimatés dans des bassins ou des réservoirs où l'eau peut s'écouler, à l'abri de toute perturbation physique et, de préférence, dans un endroit séparé des réservoirs d'essai.

Après livraison aux installations d'élevage ou d'acclimatation du laboratoire chargé des essais, les poissons doivent être détenus pour au moins deux semaines dans les conditions précisées à la section 2.4. Cette période d'acclimatation doit immédiatement précéder l'essai auquel ils vont être soumis. L'acclimatation des poissons peut se dérouler à l'intérieur ou à l'extérieur, dans des réservoirs dotés de couvercles munis d'appareils d'éclairage à réglage photopériodique. La publication d'Environnement Canada (EC, 1990a) donne de plus amples détails sur la détention et l'acclimatation des poissons devant être soumis à des essais de toxicité.

Les réservoirs devraient rester propres, la nourriture excédentaire et les excréments étant siphonnés aussi souvent que nécessaire. Il est recommandé d'utiliser des réservoirs dotés de colonnes d'alimentation doubles et centrales, qui assurent un autonettoyage partiel. Il faudrit désinfecter les réservoirs et les rincer à fond avec de l'eau ayant servi à la détention ou à l'acclimatation des poissons avant d'y déposer un nouveau lot de

poissons. On devrait utiliser des désinfectants renfermant des composés chlorés ou iodophores ou du chlorure de n-alkyldiméthylbenzylammonium.

Sauf indication contraire du fabricant, il faudrait nourrir les poissons une ou plusieurs fois par jour avec une nourriture en boulettes commerciale (standard) reconnue, la ration quotidienne variant entre 1 et 5 % du poids frais des poissons, selon la température de l'eau et leur taille (EC, 1990a). Le type et la taille des boulettes devraient être choisis en fonction de la taille et de l'âge des poissons, de la température de l'eau et des recommandations du fabricant. On devrait également suivre les recommandations du fabricant en ce qui touche la durée et la méthode de stockage de la nourriture.

Les poissons morts ou moribonds devraient être enlevés immédiatement après l'inspection quotidienne.

Les mortalités dans les réservoirs d'où proviendront les poissons soumis à l'essai devraient être surveillées et consignées quotidiennement, et doivent l'être au minimum cinq jours par semaine. Pendant la période de sept jours précédant le jour de mise en route de l'essai, le taux cumulatif de mortalité des poissons doit être inférieur à 2 %. Si le taux cumulatif de mortalité durant cette période est de 2 à 10 %, l'acclimatation doit être prolongée d'une période additionnelle d'au moins sept jours, et jusqu'à ce que le taux cumulatif de mortalité soit inférieur à 2 % pour la période de sept jours précédant la mise en route de l'essai. Si le taux de mortalité dépasse 10 % par semaine, pour une période quelconque de sept jours, le groupe de poissons est inadmissible pour utilisation ultérieure lorsque les mortalités sont causées par la maladie ou la présence de contaminants aquatiques; lorsqu'elles sont attribuables à d'autres facteurs (p. ex., taux de mortalité initial élevé pendant le passage au stade de l'alimentation active ou à la suite du transfert des poissons), on peut utiliser les poissons pour des essais de toxicité à condition que le taux de mortalité dans les réservoirs d'où ils proviennent baisse à moins de 2 % pendant les sept jours précédant immédiatement le jour de mise en route de l'essai.

Il faudrait éviter de traiter les poissons malades à l'aide de produits chimiques. Si un tel traitement est inévitable, il faut attendre au moins deux semaines avant de pouvoir les utiliser pour des essais. L'essai portant sur un produit toxique de référence (section 7) permet d'évaluer dans une certaine mesure si les poissons conviennent pour des essais de toxicité.

2.3 Eau

L'eau pour la détention et l'acclimatation des poissons peut être de l'eau souterraine ou de surface non contaminée ou encore de l'eau potable déchlorée provenant du réseau municipal d'alimentation. Elle devrait favoriser de façon constante la survie, la santé et le développement de la truite arc-en-ciel. On devrait évaluer la qualité chimique de l'eau du laboratoire aussi souvent que nécessaire pour en consigner les variations. On devrait mesurer au moins la dureté, le pH, la conductivité, l'oxygène dissous et le chlore résiduel total (s'il s'agit d'eau potable provenant du réseau municipal d'alimentation) et, au besoin, l'alcalinité, les matières solides en suspension, le carbone organique total, les gaz totaux dissous, l'ammoniac, les nitrites, les métaux et

l'ensemble des pesticides organophosphorés. Toute sursaturation en gaz devrait être éliminée (voir EC, 1990a).

Si l'on utilise de l'eau déchlorée provenant du réseau municipal d'alimentation, on doit voir à ce qu'elle soit exempte de toute concentration nocive de chlore ou de composé chlorés au moment de l'exposition des poissons. L'objectif établi pour la concentration de chlore résiduel total dans les réservoirs de détention et dans l'eau de contrôle/de dilution des réservoirs d'essai est d'au plus 0,002 mg/L (voir EC, 1990a).

Le débit d'eau fraîche dans les réservoirs de détention et d'acclimatation doit être ≥1 L/min par kilogramme de poissons détenus (soit 1,4 L/g poisson · d ou 0,69 g poisson · d/L). De plus, les réservoirs doivent renfermer, en tout temps, ≥1 L d'eau par 10 g de poissons détenus. Le débit de l'eau dans chaque réservoir devrait être mesuré et consigné périodiquement (p. ex., quotidiennement ou, au minimum, hebdomadairement). De plus, il faudrait périodiquement (p. ex., chaque semaine) prélever au hasard dix poissons ou plus de chaque réservoir d'acclimatation/de détention pour déterminer les poids frais individuels et s'assurer que les exigences stipulées sont respectées. Le poids frais moyen individuel des poissons prélevés devrait être déterminé et consigné pour chacun de ces échantillons. Ces mesures devraient aussi servir d'indicateur lorsqu'on détermine le volume d'effluent nécessaire pour un essai et pour s'assurer de ne pas dépasser la densité maximale de 0,5 g de poisson/litre de solution dans chaque récipient d'essai au cours de l'essai de toxicité (voir section 4.2).

2.4 Conditions physico-chimiques

L'éclairement devrait être assuré par des appareils fluorescents à spectre continu et avoir une intensité comprise entre 100 et 500 lux à la surface de l'eau. Pendant au moins deux semaines avant l'essai, on doit maintenir une photopériode constante de 16 ± 1 h de lumière et 8 ± 1 h d'obscurité, préférablement avec une période de transition de 15 à 30 minutes.

La température de détention peut se situer entre 4 et 18° C, mais les poissons doivent être acclimatés pendant au moins deux semaines et, de préférence, pendant trois semaines ou plus, à une température de 15 ± 2° C avant d'être soumis à un essai. La température de l'eau peut être modifiée à raison d'au plus 3° C/d. La teneur en oxygène dissous de l'eau des réservoirs devrait être de 80 à 100 % de saturation en air. Au besoin, on devrait assurer une aération supplémentaire des réservoirs au moyen d'air comprimé filtré et exempt d'huile. Le pH de l'eau devrait être compris entre 6,0 et 8,5. On devrait contrôler, de préférence tous les jours, la température de l'eau de chaque bassin d'acclimatation ou de détention ainsi que sa teneur en oxygène dissous, son pH, son débit et le nombre de poissons morts. Il est aussi recommandé de contrôler chaque semaine, ou plus fréquemment, la teneur en ammoniac et en nitrites dans les réservoirs d'acclimatation/de détention. Si l'on utilise de l'eau déchlorée du réseau municipal d'alimentation, il est de plus recommandé de mesurer une fois par semaine, ou plus fréquemment, la teneur en chlore résiduel total dans ces réservoirs.

Installations

L'essai doit être réalisé dans une installation isolée des allées et venues générales du laboratoire, soit dans une pièce isolée ou dans une zone entourée d'un mur ou d'un rideau. La poussière et les émanations devraient être réduites au minimum. On peut maintenir la température d'essai (15 ± 1 ° C) au moyen d'un appareil de conditionnement de l'air réglé par thermostat ou en immergeant les réservoirs d'essai dans des bains-marie à température contrôlée.

Les réservoirs d'essai et tout l'équipement pouvant entrer en contact avec des solutions d'essai ou l'eau de contrôle/de dilution ne doivent pas contenir de substances pouvant être entraînées dans les solutions d'essai ou absorber les produits toxiques qu'elles contiennent. Les réservoirs d'essai doivent être en verre ou en Plexiglas, en acrylique, en polypropylène, en polyéthylène, ou être revêtus de polyéthylène (non toxique). Si l'on utilise des revêtements, il faut les jeter à la fin de l'essai. Il est recommandé de couvrir les réservoirs d'essai au besoin pour éviter que les poissons ne s'en échappent (EC, 1990a). Tous les réservoirs utilisés pour un essai devraient être identiques et contenir au moins 15 cm d'eau. L'équipement doit être nettoyé et rincé à fond conformément à de bonnes pratiques de laboratoire.

L'eau de contrôle/de dilution devrait correspondre à la description fournie à la section 2.3 et, de préférence, être identique à l'eau utilisée pour la détention et ou l'acclimatation des poissons.

Mode opératoire général pour la détermination de la létalité aiguë d'effluents

4.1 *Étiquetage, transport et stockage des échantillons*

Le volume requis pour les échantillons dépend de la taille des poissons, du nombre de poissons par solution d'essai, des exigences concernant la densité de chargement, des concentrations d'essai et de l'utilisation de répétitions. Normalement, le volume nécessaire est de 25 à 50 L ou plus pour les essais à concentration unique et de 50 à 100 L ou plus pour ceux visant à établir une CL50.

Les récipients utilisés pour le transport et le stockage des échantillons doivent être fabriqués de matériaux non toxiques (p. ex., bonbonnes ou seaux de polyéthylène ou de polypropylène). Ils doivent être neufs, ou encore nettoyés à fond et séchés, et on devrait les rincer avec de l'eau propre, puis avec l'échantillon à recueillir. Il faudrait les remplir afin de réduire les vides d'air, puis les fermer hermétiquement (p. ex., avec un couvercle-pression si le récipient utilisé est un seau). L'étiquette qu'ils portent doit indiquer au moins le type d'échantillon, la source, la date et l'heure du prélèvement ainsi que le nom du (des) préposé(s) à l'échantillonnage.

Les échantillons ne doivent pas geler. Pendant le transport, ils devraient être conservés dans l'obscurité, à une température de 1 à 8° C si le voyage dure plus de deux jours. À la réception du ou des échantillon(s) au laboratoire, il faudrait mesurer et consigner la température de l'effluent dans chaque récipient contenant les échantillons. La température de la portion de chaque échantillon destinée à l'essai de toxicité doit être ajustée à 15 ± 1° C, avant d'entreprendre l'essai de toxicité.

Pour permettre la mise en route de l'essai le jour de l'arrivée de l'échantillon au laboratoire, on peut ajuster rapidement la température de l'échantillon ou des échantillons d'effluent (voir section 4.3). Une autre option consiste à entreposer l'échantillon à l'obscurité à 4 ± 2° C pendant une brève période (p. ex., pendant la fin de semaine si les échantillons sont reçus le vendredi après-midi). Si l'on a recours à cette alternative, l'échantillon doit être entreposé dans des récipients remplis à capacité et hermétiquement fermés, et gardés à l'obscurité dans une installation réfrigérée. Une troisième option est de garder l'échantillon pendant la nuit dans une installation dont la température est ajustée à celle de la température d'essai (c'est-à-dire, 15 ± 1° C); dans cette circonstance, l'essai peut être entrepris le lendemain. Si un échantillon est réchauffé ou refroidi à 15 ± 1° C au cours de la nuit, il faut dans ce cas le garder dans un ou plusieurs récipients remplis à capacité et hermétiquement fermés.

L'essai des échantillons devrait être entrepris le plus tôt possible après leur prélèvement. Il devrait débuter dans les trois jours et doit être commencé au plus tard cinq jours après la fin des prélèvements. Le contenu de chaque récipient doit être agité vigoureusement just avant de verser des aliquotes pour préparer les solutions. Les

sous-échantillons (c.-à-d. les aliquotes réparties dans deux ou plusieurs récipients) doivent être combinés.

4.2 Conditions de l'essai

Il s'agit d'un essai statique, c'est-à-dire sans renouvellement des solutions, d'une durée de 96 h. La densité de chargement de poissons dans chaque réservoir d'essai ne doit pas dépasser 0,5 g/L. Le respect de cette exigence est basé sur le poids frais moyen des poissons témoins à la fin de l'essai (section 4.5). Il ne faut pas nourrir les poissons pendant l'essai ni pendant les 16 h qui le précèdent. L'essai n'est pas valable si plus de 10 % des poissons témoins meurent ou manifestent un comportement atypique ou stressé (EC, 1990a).

L'essai doit être mené à 15 ± 1 °C. Toutes les solutions doivent être aérées pendant l'essai à un débit contrôlé de 6,5 ± 1 mL/min · L. L'éclairement et la photopériode (16 ± 1 h de lumière, 8 ± 1 h d'obscurité) doivent être les mêmes que les conditions définies pour l'acclimatation (section 2.4).

L'essai doit être réalisé sans correction du pH de l'échantillon ou des solutions. Toutefois, si l'on désire comprendre dans quelle mesure les valeurs extrêmes du pH (c.-à-d. les valeurs inférieures à 5,5 ou supérieures à 8,5) peuvent contribuer à la létalité aiguë de l'échantillon ou des solutions, on peut effectuer parallèlement un deuxième essai (avec correction du pH). Si deux essais sont ainsi menés, les résultats définitifs devraient être ceux obtenus à la suite de l'essai sans correction du pH. Environnement Canada (1990a) fournit des explications et des détails concernant la correction du pH. L'ajustement du pH est également l'une de techniques d'approche expérimentale qui permettent de définir la cause de la toxicité des échantillons (USEPA, 1991).

4.3 Préparation des solutions d'essai

La température de l'échantillon d'effluent et de l'eau de contrôle/de dilution doit être ramenée à 15 ± 1° C si elle ne se situe pas dans cet intervalle. L'échantillon peut être refroidi au moyen d'un bain d'eau froide ou d'un refroidisseur à immersion fait d'un matériau non toxique. Pour le réchauffer, utiliser un bain-marie. Ne *pas utiliser* de thermoplongeur pour réchauffer les échantillons ou les solutions d'essai.

Pour un essai donné, il faut utiliser la même eau pour préparer la ou les solutions de contrôle et toutes les concentrations d'essai inférieures à 100 %. Il s'agit presque toujours de la même eau qui a servi à l'acclimatation. Si l'on en augmente la température, il faut éviter la sursaturation en gaz. L'eau doit avoir une teneur en oxygène de 90 à 100 % de saturation en air, obtenue au besoin grâce à une aération vigoureuse au moyen d'air comprimé exempt d'huile passant par des pierres de barbotage ou un diffuseur en verre.

Les réservoirs d'essai devraient être rincés avec de l'eau de contrôle/de dilution juste avant d'être utilisés, mais cela n'est pas nécessaire si l'on a recours à des revêtements de polyéthylène jetables. Chaque solution d'essai doit avoir le même volume et être bien mélangée avec une baguette de verre, un agitateur en téflon ou un autre dispositif fait d'un matériau non toxique, juste avant l'utilisation. Tous les réservoirs d'essai, dispositifs de mesure, appareils d'agitation et

seaux pour le transfert des poissons doivent être nettoyés et rincés à fond, conformément aux modes opératoires normalisés.

Chaque réservoir d'essai doit contenir au moins 15 cm de solution. Une fois préparée, chaque solution d'essai doit être aérée pendant 30 minutes à raison de 6,5 ± 1 mL/min · L. Par la suite, on doit mesurer la teneur en oxygène dissous au moins dans la solution de la concentration la plus élevée (normalement l'échantillon non dilué). Si, et seulement si, cette valeur est inférieure à 70 % ou supérieure à 100 % de saturation en air, il faut alors, avant l'exposition des poissons, continuer d'aérer toutes les solutions d'essai, y compris les solutions de contrôle, à raison de 6,5 ± 1 mL/L · min. Il ne faut pas poursuivre l'aération au-delà de 90 minutes additionnelles. Si une période d'aération plus courte a permis d'obtenir 70 % de saturation en air dans la solution d'essai de la concentration la plus élevée (ou 100 % de saturation, s'il y avait sursaturation), l'aération est arrêtée au temps requis pour atteindre ces valeurs. Immédiatement après, on doit introduire les poissons dans chaque solution d'essai et mettre en route l'essai, qu'on ait obtenu ou non de 70 à 100 % de saturation dans toutes les solutions d'essai. L'aération devrait être assurée par des bulles d'air comprimé passant par un diffuseur d'air propre en verre de silice ou par une pipette de verre jetable (EC, 1990a). L'aération de chaque solution d'essai doit être maintenue tout au long de l'essai à raison de 6,5 ± 1 mL/L · min.

4.4 Mise en route de l'essai

Il faut préparer une ou plusieurs solutions de contrôle et les inclure dans chaque essai conduit avec chaque échantillon.

L'utilisation d'une seule soution de contrôle et des poissons qu'elle contient pour plus d'un essai de toxicité et (ou) pour plus d'un échantillon d'effluent est inacceptable.

Chaque récipient d'essai doit porter un code ou une étiquette précisant clairement la concentration de la solution ainsi que la date et l'heure du début de l'essai. Les récipients d'essai devraient être installés de façon à faciliter l'observation des poissons. Les concentrations à expérimenter devraient être placées selon un ordre aléatoire s'il s'agit d'un essai à concentrations multiples (section 6).

Les poissons en bonne santé qui ont été acclimatés pendant au moins deux semaines aux conditions de température et d'éclairement utilisées pour l'essai (voir sections 2.1, 2.2, 2.3, et 2.4), et qui sont destinés à l'essai, doivent être prélevés au hasard du (des) bassins(s) d'acclimatation. Les méthodes de manipulation et de transfert devraient minimiser le stress (EC, 1990a).

On doit introduire au moins dix poissons dans chaque concentration d'essai, y compris chaque solution de contrôle. Les poissons peuvent être répartis dans deux récipients d'essai ou plus de la même concentration, de façon à respecter la densité de chargement requise (voir section 4.2).

En plus de placer les récipients d'essai selon un ordre aléatoire dans l'aire d'expérimentation, il faudrait également introduire les poissons dans chaque solution d'essai selon un ordre aléatoire. Un ou deux poissons devraient être introduits tour à tour dans chaque solution d'essai, y compris la ou les solution(s) de contrôle, jusqu'à ce que dix poissons aient été introduits dans chaque récipient.

Si une ou plusieurs solutions d'essai sont fortement colorées, opaques ou mousseuses, on peut utiliser des paniers d'un matériau non toxique et non abrasif (p. ex., nylon, polyéthylène ou polypropylène) pour inspecter les poissons pendant l'essai. On devrait alors placer un de ces paniers dans chacun des récipients d'essai (y compris le ou les contrôle[s]). Avant usage, chaque panier doit être nettoyé et rincé à fond avec de l'eau de contrôle/de dilution.

4.5 Observations et mesures

On devrait noter au début de l'essai la couleur, la turbidité et l'odeur de l'échantillon, ainsi que la présence de matières solides flottantes ou décantables. Il faudrait aussi observer l'apparence des solutions d'essai, et consigner les modifications survenant au cours de l'essai.

La teneur en oxygène dissous, le pH et la température de chaque solution d'essai, y compris la ou les solutions de contrôle, doivent être mesurés au moins au début et à la fin de l'essai. Les dernières mesures devraient être effectuées une fois les observations biologiques terminées. La conductivité de chaque solution d'essai doit être mesurée au moins au début de l'essai.

Il faut procéder fréquemment à l'observation de routine des poissons dans chaque réservoir d'essai afin de préciser à quel moment la mort est survenue, identifier tout effet toxique sublétal évident, et retirer les poissons morts qui autrement pourraient souiller la solution d'essai. Dans la mesure du possible, chaque réservoir d'essai devrait être inspecté au moins 24, 48, 72 et 96 h après le début de l'essai. Des observations plus fréquentes sont indiquées au cours du premier jour. Les poissons morts doivent être consignés et retirés lors de chaque inspection. Les poissons sont réputés morts quand ils ne laissent apparaître aucun signe d'activité operculaire ou autre et qu'ils ne réagissent pas à une légère poussée. Il faudrait également consigner les effets toxiques sublétaux évidents (EC, 1990a). Si les solutions sont fortement colorées, opaques ou mousseuses, on peut inspecter les poissons au moyen d'une épuisette (nettoyée et rincée avant usage) ou en les soulevant à la surface au moyen d'un panier approprié (section 4.4).

On doit déterminer et consigner, à la fin de l'essai, la longueur moyenne à la fourche des poissons témoins. Les mesures de la longueur à la fourche, pour chaque poisson témoin, devraient servir à évaluer si la taille des poissons soumis à l'essai correspond à la recommandation à l'effet que *"le plus gros poisson ne devrait pas être plus de deux fois plus long que le plus petit poisson soumis au même essai"* (section 2.1).

Le poids frais moyen de chaque poisson témoin doit être déterminé et consigné à la fin de l'essai. Cette mesure doit servir à confimer que l'on a respecté la gamme de poids acceptable pour les poissons soumis à l'essai, précisée à la section 2.1 (c.-à-d., *"leur poids frais moyen doit se situer entre 0,3 et 2,5 g"*). Il doit aussi servir à confirmer que l'on a respecté la densité de chargement des poissons dans les soution d'essai, précisée à la section 4.2 (c.-à-d., *"la densité de poissons dans chaque récipient d'essai ne doit pas dépasser 0,5 g/L"*).

Tous les poissons encore en vie (y compris les témoins) à la fin de l'essai doivent être éliminés de manière humanitaire. On recommande à cette fin une surdose d'un agent anesthésique comme le méthosulfonate de tricaïne.

Section 5

Mode opératoire pour la détermination du taux de mortalité après 96 h dans un essai à concentration unique

Toutes les conditions, méthodes et installations prévues aux sections 1, 2, 3, 4, 7 et 8 s'appliquent au présent mode opératoire.

L'essai porte sur une seule concentration d'effluent, soit l'effluent non dilué, sauf indication contraire, et sur une solution de contrôle. Pour cet essai, on recommande des répétitions (p. ex., trois répétitions de l'effluent non dilué et trois répétitions des solutions de contrôle, en utilisant dix poissons par solution répétée), afin d'avoir une plus grande confiance dans les résultats finaux et leur interprétation. Au moins dix poissons doivent être exposés à chaque solution d'essai et de contrôle. L'essai n'est pas valide si plus de 10 % des poissons témoins (données combinées si l'on utilise des solutions répétées) meurent ou laissent apparaître un comportement atypique ou stressé.

Le résultat final de cet essai est le pourcentage de mortalité après 96 h. On utilise couramment une valeur de 50 %. Par exemple, en vertu du Règlement fédéral de 1992 sur les effluents des fabriques de pâtes et papiers découlant de la *Loi sur les pêches* (Gouvernement du Canada, 1992), un effluent n'est pas acceptable si 50 % ou plus des poissons meurent après avoir été exposés à une concentration de 100 %.

Mode opératoire pour la détermination de la CL50–96 h dans un essai à concentrations multiples

Toutes les conditions, méthodes et installations prévues aux sections 1, 2, 3, 4, 7 et 8 s'appliquent au présent mode opératoire.

Pour déterminer la CL50, il faut utiliser au moins cinq concentrations d'effluent et une solution de contrôle (entièrement composée d'eau de dilution). On doit exposer au moins dix poissons à chaque concentration d'essai, y compris la concentration de 100 % et la solution de contrôle. La concentration la plus élevée doit être l'effluent non dilué (100 %), et chaque concentration successive doit correspondre à au moins 50 % de la précédente. Il est avantageux d'utiliser une série géométrique (ou logarithmique), par exemple des concentrations de 100, 50, 25, 12,5 et 6,3. Les concentrations peuvent être basées sur d'autres proportions ou sur des séries de dilutions normalisées (EC, 1990a, annexe D).

Du fait que cette CL50 doit inclure l'effluent non dilué comme la concentration la plus élevée, les résultats de cet essai peuvent aussi servir à déterminer le paramètre d'un essai avec une concentration unique, soit le pourcentage final de mortalité après 96 h dans l'effluent non dilué (voir section 5).

On peut prévoir des répétitions de chaque concentration, mais cela n'est pas une exigence. Si on le fait, les données obtenues doivent être combinées pour le calcul de la CL50. La précision de l'estimation augmente avec le nombre de poissons exposés à chaque concentration d'essai, sans nécessairement en améliorer l'exactitude.

On devrait calculer la CL50 après 96 h et ses limites de confiance à 95 % et indiquer la méthode utilisée pour ces calculs. Des programmes informatiques existent à cette fin (EC, 1990a), et on devrait y avoir recours. Un programme particulier est recommandé; les personnes qui fournissent une disquette peuvent se le procurer auprès d'Environnement Canada (adresses à l'annexe), avec la permission de C.E. Stephan (1977). On devrait vérifier toute CL50 calculée par ordinateur en examinant, sur une échelle de probabilité logarithmique, la courbe des pourcentages de mortalité après 96 h pour les diverses concentrations d'effluent (EC, 1990a).

L'essai n'est pas valide (les données étant combinées s'il y a des répétitions) plus de 10 % des poissons témoins meurent ou laissent apparaître un comportement atypique ou stressé.

Mode opératoire pour les essais avec un produit toxique de référence

On doit utiliser un produit toxique de référence pour évaluer la sensibilité relative de la population de poissons utilisée dans l'essai de toxicité ainsi que la précision et la fiabilité des données produites par le personnel du laboratoire pour ce produit toxique de référence dans des conditions d'essai normalisées. Le ou les produits chimiques de référence en cause doivent faire l'objet d'un essai au moins une fois par mois civil où l'on procède à des essais d'effluents, et au moment de l'acclimatation d'un nouveau lot de poissons. Les poissons utilisés pour l'essai du ou des produit(s) toxique(s) de référence mené conjointement avec un essai servant à déterminer la létalité aiguë d'un effluent doivent provenir du même groupe que ceux utilisés pour l'essai conduit avec l'effluent. Les méthodes et les conditions à respecter sont identiques à celles énoncées à la section 4 et déterminées par Environnement Canada (1990a, c), à ceci près que c'est un produit chimique de référence, et non un effluent, qui est soumis à l'essai. On doit utiliser pour ces essais l'eau de contrôle/de dilution qui sert couramment aux essais d'effluent.

On recommande d'utiliser comme produits toxiques de référence du phénol ou du sulfate de zinc ($ZnSO_4 \cdot 7H_2O$) de qualité «reactif». La CL50 après 96 h devrait être déterminée pour le ou les produits toxiques de référence utilisés et exprimée sous forme de mg/L de phénol ou de zinc (Zn^{++}) (EC, 1990a). Les solutions mères de phénol doivent être preparées le jour même de leur utilisation; les solutions de zinc devraient être préparées le jour même de leur utilisation ou alors être conservées à l'obscurité à un pH entre 3 à 4.

Les concentrations du produit toxique de référence dans toutes les solutions mères devraient être mesurées au moyen de méthodes chimiques appropriées (APHA *et al.*, 1998). Au moment de la préparation des solutions d'essai, on doit prélever des aliquotes au moins dans la solution de contrôle et dans les solutions à teneur inférieure, moyenne et supérieure; ces aliquotes devraient être analysées immédiatement ou stockées pour analyse future, au cas où la CL50 serait atypique (c.-à-d., à l'extérieur des limites de contrôle). Si elles sont stockées, on doit les garder à l'obscurité, à une température de 4 ± 2° C. Un agent conservateur devrait être ajouté aux parties aliquotes de solutions de zinc et de phénol avant de les stocker (APHA *et al.*, 1998). Les aliquotes d'échantillon stockées nécessitant une mesure chimique devraient être analysées aussitôt l'essai de toxicité terminé. Il est souhaitable, mais non nécessaire, de mesurer les concentrations dans les même solutions à la fin de l'essai, une fois les observations biologiques terminées. Le calcul de la CL50 devrait être fondé sur les concentrations mesurées, si elles diffèrent de façon appréciable (c.-à-d., d'au moins 20 %) des concentrations nominales.

Une fois que l'on a recueilli suffisamment de données (EC, 1990c), on doit établir et continuellement mettre à jour un diagramme de contrôle qui présente graphiquement les CL50 pour chaque produit toxique de

référence utilisé. Ce diagramme devrait rapporter la concentration logarithmique en ordonnée et la date ou le numéro de l'essai en abscisse. Chaque nouvelle CL50 pour le produit toxique de référence devrait être comparée aux limites d'avertissement établies; pour être acceptable, la CL50 doit s'inscrire à l'intérieur des limites. La moyenne et l'écart type doivent toujours être calculés sur une échelle logarithmique. La moyenne logarithmique et ses limites d'avertissement inférieure et supérieure (± 2 fois l'écart type) calculées en prenant les valeurs logartihmiques disponibles, sont recalculées pour chaque CL50 successive, jusqu'à ce que les statistiques se stabilisent (USEPA, 1985a; EC, 1990c).

Le diagramme de contrôle peut être établi en reportant simplement la moyenne et ± 2 fois l'écart type comme logarithmes, ou si l'on préfère, en les convertissant en valeurs arithmétiques et en reportant la CL50 ± 2 fois l'écart type sur une échelle logarithmique de concentration. Si une CL_{50} s'inscrivait en dehors des limites de contrôle, la sensibilité des poissons soumis à l'essai et la performance ainsi que la précision de l'essai seraient mises en doute. Dans de telles circonstances, on doit vérifier toutes les conditions de détention et d'essai; selon les constatations faites, il se peut qu'on doive prolonger l'acclimatation des poissons et les soumettre à une nouvelle évaluation au moyen d'un ou de plusieurs produits toxiques de référence, ou encore obtenir une nouvelle population de poissons et l'acclimater avant de l'utiliser pour des essais de toxicité avec un effluent et un ou des produits toxiques de référence.

Présentation des résultats

Voici un résumé des exigences de la présente méthode de référence en matière d'établissement du procès-verbal des essais et de tenue de dossiers. Pour de plus amples détails, on peut consulter les sections précédentes de cette méthode.

Sauf indication contraire d'Environnement Canada, tous les renseignements énumérés à la section 8.1 doivent être communiqués à Environnement Canada pour chaque essai de toxicité qui est entrepris. Ces renseignements doivent être fournis conformément aux règlements pertinents et dans la forme et les moyens précisés par Environnement Canada* (c.-à-d., présentation manuelle ou informatique; mode de transmission; forme et contenu).

Environnement Canada peut aussi exiger la communication d'informations autres que celles spécifiées en section 8.1, par exemple des renseignements requis en vertu d'un règlement particulier ou nécessaires pour faciliter la compréhension ou l'évaluation des données consignées au procès-verbal.

Sauf indication contraire d'Environnement Canada, les renseignements énumérés à la section 8.2 dovient être consignés et versés au dossier pour une période de cinq ans, et on doit pouvoir les produire à la demande d'Environnement Canada. Ils seront requis de façon moins fréquente, par exemple pendant un audit ou une enquête.

* Pour obtenir des détails, prière de communiquer avec un des bureaux énumerés à l'annexe.

8.1 Données à consigner au procès-verbal

8.1.1 Effluent

- Nom et adresse de l'installation produisant l'effluent.
- Date et heure du prélèvement de l'échantillon.
- Type d'échantillon (p. ex., «effluent non traité de l'usine», «effluent terminal de l'usine», «rejet de la lagune de déversement en cas d'urgence», «lixiviat»).
- Brève description du lieu de prélèvement.
- Méthode d'échantillonnage (p. ex., «échantillon instantané», «échantillon discontinu», «échantillon composite de 24 h, avec prélèvement de sous-échantillons à toutes les heures»).
- Personne ayant prélevé l'échantillon.

8.1.2 Installations et conditions de l'essai

- Type et méthode d'essai (p. ex., «essai à concentration unique», «méthode décrite dans la deuxième édition (Décembre 2000) de la publication SPE 1/RM/13»).
- Indication de toute exigence, exprimée par «doit», «doivent» ou «il faut», stipulée dans les sections 2 à 7 de ce rapport, qui n'a pas été respectée, avec les détails relatifs au non-respect de la ou des exigences.

- Nom du laboratoire d'essai et ville où il se trouve.
- Pourcentage de mortalité des poissons dans le ou les réservoir(s) d'où proviennent les poissons soumis aux essais, tel qu'enregistré chaque jour (ou au minimum cinq jours sur les sept compris dans une période hebdomadaire), pour la période de sept jours précédant immédiatement l'essai.
- Espèce d'organisme soumis à l'essai.
- Date et heure de mise en route de l'essai défintif.
- Personne(s) ayant mené l'essai et vérifié les résultats.
- pH, température, teneur en oxygène dissous et conductivité de l'effluent non dilué et non corrigé, juste avant la préparation de solutions d'essai.
- Confirmation que le pH de l'échantillon ou de la solution d'essai n'a pas été corrigé. Indication de la méthode utilisée pour la correction du pH si des essais avec et sans correction du pH ont été conduits (voir section 4.2).
- Renseignements sur toute aération des solutions d'essai avant l'introduction des poissons (débit, durée). Débit d'aération pendant l'essai.
- Concentration et volume des solutions d'essai, y compris les solutions de contrôle, et mention de l'usage de répétitions, s'il y a lieu.
- Mesures de la teneur en oxygène dissous, du pH et de la température de chaque solution d'essai, y compris des solutions de contrôle, au début et à la fin de l'essai. Mesure de la conductivité de chaque solution d'essai au début de l'essai.
- Nombre de poissons introduits dans chaque récipient d'essai.
- Longueur moyenne à la fourche des poissons témoins, à la fin de l'essai, avec intervalles de valeurs mesurées.
- Poids frais moyen individuel des poissons témoins à la fin de l'essai.
- Densité estimée de chargement (g/L) des poissons dans les solutions d'essai.

8.1.3 Résultats

- Nombre de poissons morts dans chaque solution d'essai (y compris les solutions de contrôle) après 96 h. Nombre de poissons témoins laissant apparaître un comportement atypique ou stressé.
- Taux moyen de mortalité dans les solutions d'effluent et l'eau de contrôle, s'il s'agit d'un essai à concentration unique ou à concentrations multiples conduit en utilisant des solutions répétées. Nombre moyen de poissons témoins montrant un comportement atypique ou stressé si on a utilisé des solutions de contrôle répétées.
- Pour les essais à concentrations multiples, estimation de la CL50 et ses limites de confiance à 95 %, si statistiquement possibles. Indication de la méthode statistique utilisée (p. ex., «probabilité logarithmique», «moyenne mobile»).
- Dernière valeur de la CL50 après 96 h (et de ses limites de confiance à 95 %)

calculée pour le(s) essai(s) avec le ou les produit(s) toxique(s) de référence conduit(s) avec le groupe de poissons utilisés dans les essais avec l'effluent. Produit(s) chimique(s) de référence. Date de mise en route de l'essai. Moyenne géométrique des CL50 calculées antérieurement et limites d'avertissement (± 2 fois l'écart type).

8.2 Données à verser au dossier**

8.2.1 Effluent

- Tout renseignement (p. ex., code, description de l'échantillon, date/heure d'échantillonnage) inscrit sur la ou les étiquettes du ou des récipients contenant l'échantillon.

- Volume de l'échantillon.

- Conditions de transport et de stockage à l'arrivée au laboratoire et durant l'entreposage. Mention si l'échantillon était gelé en tout ou en partie à l'arrivée.

- Apparence et autres propriétés (observations sur la couleur, la turbidité, l'odeur et la présence de solides flottants ou décantables).

- Modification de la couleur, précipitation, floculation, rejet de substances volatiles et autres modifications survenues pendant la préparation des solutions d'essai.

- Mode opératoire et résultats de toute analyse chimique (p. ex., teneur en solides en suspension, dureté) effectuée sur l'effluent, si disponibles.

8.2.2 Installations et conditions de l'essai

- Adresse du laboratoire d'essai.

- Description des installations d'élevage et d'essai, y compris leur agencement général et leurs moyens d'isolation.

- Conditions normales de détention et d'acclimatation : récipients, emplacement, éclairement, températures y compris le taux maximal de variation, aération, volume et débit de l'eau, mode utilisé pour le renouvellement de l'eau, nombre et densité des poissons, mode de manipulation, type de nourriture, ration et fréquence de l'alimentation, incidence et traitement éventuel de maladies, pourcentage hebdomadaire cumulatif de mortalité.

- Provenance des poissons soumis à l'essai.

- Bref résumé des conditions et modes opératoires propres à l'essai en ce qui touche la détention et l'acclimatation des poissons, s'ils diffèrent de la pratique courante (p. ex., durée, source et caractéristiques de l'eau employée telles que la température, le pH et la teneur en oxygène dissous, type de nourriture et ration, incidence et traitement de maladies, pourcentage hebdomadaire de mortalité).

- Desciption de la ou des source(s) de l'eau utilisée pour l'élevage et l'acclimatation des poissons et comme eau de contrôle/de dilution.

** Ces données dovient être conservées pendant cinq ans à l'installation d'essai ou dans les bureaux du responsable des rejets ou encore à ces deux endroits. Certaines d'entre elles peuvent s'appliquer à une série d'essais et être consignées et conservées dans un rapport général.

- Prétraitement de l'eau d'acclimatation et de l'eau de contrôle/de dilution, le cas échéant (p. ex., correction de la température, débit et durée de l'aération, quantité de tout produit chimique ajouté).

- Mesures (moyennes et intervalles) de la qualité de l'eau d'acclimatation et de l'eau de contrôle/de dilution effectuées sur l'approvisionnement en eau et dans les bassins de détention. Renseignements à fournir : dureté, pH, conductivité, teneur en oxygène dissous et chlore résiduel total (s'il s'agit d'eau déchlorée provenant du réseau d'alimentation municipal). Autre renseignements également souhaitables : gaz totaux dissous, alcalinité, matières solides, carbone organique, couleur, ions minéraux, métaux, ammoniac, nitrites et pesticides organophosphorés.

- Systèmes de réglage de l'éclairement et de la température.

- Source d'éclairement, photopériode et mesures antérieures de l'éclairement dans les bassins d'élevage ou d'acclimatation et à la surface des réservoirs d'essai.

- Description des réservoirs d'essai (dimension, forme et matériau), des couvercles, des paniers (si on utilise pour inspecter les poissons) et leurs méthodes courantes de nettoyage.

- Méthode utilisée pour l'introduction selon un ordre aléatoire des poissons dans les récipient d'essai.

- Méthodes et appareillage pour l'aération des solutions d'essai.

- Toutes les mesures individuelles de longueur à la fourche et du poids frais des poissons soumis à l'essai (avec les valeurs moyennes et la gamme de valeurs pour chaque échantillon).

- Hauteur de solution d'essai dans les récipients d'essai. Apparence des solutions, y compris toute modification évidente survenue pendant l'essai.

- Concentrations d'essai des produits toxiques de référence, tant nominales que mesurées. Indication de l'ensemble de données utilisé pour estimer la CL50.

- Toute mesure de la qualité de l'eau des solutions d'essai non consignée au procès-verbal (section 8.1.2).

8.2.3 Résultats

- Observations du nombre de poissons morts non consignées au procès-verbal (p. ex., après 24, 48, et 72 h) (section 8.1.3).

- Observations du comportement et de l'apparence des poissons dans chaque solution d'essai faites au cours de l'essai.

- Toutes représentation graphique manuelle des données utilisée pour vérifier une CL50 générée par ordinateur.

Références

APHA, AWWA, et WEF (American Public Health Association, American Water Works Association et Water Environment Federation). *Standard Methods for the Examination of Water and Wastewater,* 20e éd., L. Clesceri, A. Greenber et A. Eaton (éditeurs), APHA, AWWA et WEF, Washington, DC (1998).

ASTM (American Society for Testing and Materials). *Standard Practice for Conducting Acute Toxicity Tests with Fishes, Macroinvertebrates, and Amphibians,* ASTM, Philadelphie, PA, Rapport E729-80, 25 pages (1980).

BHSC (British Health and Safety Commission). *Methods for the Determination of Ecotoxicity, Approved Code of Practice. Notification of New Substances Regulations,* BHSC, Londres, (R-U), Rapport COP 8, 13 pages (1982).

Craig, G.R., K. Flood, J. Lee, et M. Thomson. *Protocol to Determine the Acute Lethality of Liquid Effluents to Fish,* ministère de l'Environnement de l'Ontario, Rexdale, Ontario, 9 pages (1983).

EC (Environnement Canada). *Méthode d'essai biologique : essai de létalité aiguë sur la truite arc-en-ciel.* Conservation et Protection, Ottawa, Rapport SPE 1/RM/9, incluant les modifications apportées en mai 1996, 51 pages (1990a).

EC (Environnement Canada). *Méthode d'essai biologique : Méthode de référence pour la détermination de la létalité aiguë d'effluents chez la truite arc-en-ciel.* Conservation et Protection, Ottawa, Rapport SPE 1/RM/13 avec modifications apportées en mai 1996, 18 pages (1990b).

EC (Environnement Canada). *Document d'orientation sur le contrôle de la précision des essais de toxicité au moyen de produits toxiques de référence.* Conservation et Protection, Rapport SPE 1/RM/12, Ottawa, 1990 (1990c).

Gouvernement du Canada. "Loi sur les pêches : Règlements sur les effluents des fabriques de pâtes et papiers", *Gazette du Canada, Partie II, Vol. 126, No 11,* pp. 1967–2006 (1992).

Leitritz, E. et R.C. Lewis. «Trout and Salmon Culture (Hatchery Methods)», *California Fish Bulletin, No. 164,* University of California, Divsion of Agricultural Sciences, Berkley, CA, 197 pages (1976).

McGuinness, E.J., *Procedures Manual, Aquatic Bioassay Service Projects,* Alberta Environmental Centre, Vegreville, Alberta, Rapport AECV-82M1, 56 pages (1982).

MEO (ministère de l'Environnement de l'Ontario). *Protocol for Single Concentraiton Acute Lethality Tests on Liquid Effluent Samples using Rainbow Trout,* Rexdale, Ontario 10 pages, (1989).

OCDE (Organisation de coopération et de développement économiques). *Lignes directrices de l'OCDE pour les essais de produits chimiques.* Paris, France, Document no. 203, 12 pages. (1984).

Rocchini, R.J., M.J.R. Clark, A.J. Jordan, S. Horvath, D.J. McLeay, J.A. Servizi, A. Sholund, H.J. Singleton, R.G. Watts, and R.H. Young. *Provincial Guidelines and Laboratory Procedures for Measuring Acute Lethal Toxicity of Liquid Effluents to Fish,* ministère de l'Environnement de la Colombie-Britannique, Victoria, C.-B., 18 pages (1982).

SPE (Service de la protection de l'environnement). *Règlement sur les effluents des fabriques de pâtes et papiers.* Rapport SPE 1-WP-72-1, Ottawa (1971).

SPE (Service de la protection de l'environnement). *Règles d'emploi et d'admissibilité des dispersants pour traiter les nappes de pétrole.* Rapport SPE 1-DIU-73-1, Ottawa (1973).

SPE (Service de la protection de l'environnement). *Règlement et directives sur les effluents des raffineries de pétrole.* Rapport SPE 1-WP-74-1, Ottawa, Ontario (1974).

SPE (Service de la protection de l'environnement). *Règlement et directives sur les effluents liquides des mines de métaux.* Rapport SPE 1-WP-77-1, Ottawa, Ontario (1977a).

SPE (Service de la protection de l'environnement). *Règlement et directives sur les effluents liquides de l'industrie de la viande et de la volaille.* Rapport SPE 1-WP-77-2, Ottawa, Ontario (1977b).

SPE (Service de la protection de l'environnement). *Règlement et lignes directrices sur les effluents des établissements de transfomation de la pomme de terre.* Rapport SPE 1-WP-77-4, Ottawa, Ontario (1977c).

SPE (Service de la protection de l'environnement). *Méthode normalisée de contrôle de la toxicité aiguë des effluents.* Rapport SPE 1-WP-80-1, Ottawa, Ontario (1980).

SPE (Service de la protection de l'environnement). *Lignes directrices concernant l'homologation des dispersants et leur utilisation pour traiter les nappes de pérole.* 2e éd. Rapport SPE 1-EP-84-1, Ottawa (1984).

Stephan, C.E., «Methods for Calculating an LC50», Dans *Aquatic Toxicology and Hazard Evaluation,* F.L. Mayer et J.L. Hamelink (éd.), pp. 65-84, American Society for Testing and Materials, Philadelphie, PA (1977).

UKWRC (United Kingdom Water Research Centre), *Acute Toxicity to Fish. Determination of the 96-hour LC50 of Test Substances to the Rainbow Trout Under Flow-through Conditions,* Water Research Center, Medmenham (R.-U.) Standard Operating Procedure, No. 120 02, 19 pages (1983).

USEPA (United States Environmental Protection Agency). *Methods for Measuring the Acute Toxicity of Effluents to Freshwater and Marine Organisms,* W.H. Peltier et C.I. Weber (éd.), USEPA, Cincinnati, Rapport EPA/600/4-85-013, 216 pages (1985a).

USEPA (United States Environmental Protection Agency), *Acute Toxicity Test for Freshwater Fish. Standard Evaluation Procedure,* USEPA, Hazard Evaluation Division, Washington, DC, Rapport EPA-540/9-85-006, 12 pages (1985b).

USEPA (United States Environmental Protection Agency). *Methods for Aquatic Toxicity Identification Evaluations, Phase I. Toxicity Characterization Procedures,* 2e éd., Duluth, MN, Rapport EPA/600/6-91/003 (1991).

Membres du Groupe intergouvernemental de la toxicité environnementale et adresses de l'administration centrale et des bureau régionaux d'Environnement Canada

Membres du Groupe intergouvernemental de la toxicité environnementale (au mois de décembre 2000)

Gouvernement féderal (Environnement Canada)

C. Blaise
Centre Saint-Laurent
Montréal (Québec)

M. Bombardier
Centre de technologie environnementale
Ottawa (Ontario)

U. Borgmann
Institut national de recherche sur les eaux
Burlington (Ontario)

J. Bruno
Pacific Environmental Science Centre
North Vancouver (Colombie-Britannique)

C. Buday
Pacific Environmental Science Centre
North Vancouver (Colombie-Britannique)

T. Corbin
Atlantic Environmental Science Centre
Moncton (Nouveau-Brunswick)

K. Doe
Atlantic Environmental Science Centre
Moncton (Nouveau-Brunswick)

G. Elliot
Service de la protection de l'environnement
Edmonton (Alberta)

M. Fennell
Pacific Environmental Science Centre
North Vancouver (Colombie-Britannique)

F. Gagné
Centre Saint-Laurent
Montréal (Québec)

M. Harwood
Centre Saint-Laurent
Montréal (Québec)

P. Jackman
Atlantic Environmental Science Centre
Moncton (Nouveau-Brunswick)

N. Kruper
Service de la protection de l'environnement
Edmonton (Alberta)

D. MacGregor
Centre de technologie environnementale
Ottawa (Ontario)

D. Moul
Pacific Environmental Science Centre
North Vancouver (Colombie-Britannique)

W. R. Parker
Service de la protection de l'environnement
Fredericton (Nouveau-Brunswick)

L. Porebski
Division du milieu marin
Hull (Québec)

D. Rodrigue
Direction générale de l'avancement des technologies environnementales
Hull (Québec)

R. Scroggins
Centre de technologie environnementale
Ottawa (Ontario)

A Steenkamer
Centre de technologie environnementale
Ottawa (Ontario)

G. van Aggelen (Chairperson)
Pacific Environmental Science Centre
North Vancouver (Colombie-Britannique)

R. Watts
Pacific Environmental Science Centre
North Vancouver (Colombie-Britannique)

P. Wells
Service de la conservation de l'environnement
Dartmouth (Nouvelle-Écosse)

Gouvernement fédéral

(Pêches et Océans Canada)

R. Roy
Institut Maurice-Lamontagne
Mont-Joli (Québec)

Gouvernements provinciaux

C. Bastien
ministère de l'Environnement du Québec
Ste-Foy (Québec)

B. Bayer
ministère de l'Environnement du Manitoba
Winnipeg (Manitoba)

D. Bedard
ministère de l'Environnement de l'Ontario
Rexdale (Ontario)

M. Mueller
ministère de l'Environnement de l'Ontario
Rexdale (Ontario)

D. Poirier
ministère de l'Environnement de l'Ontario
Rexdale (Ontario)

J. Schroeder
ministère de l'Environnement de l'Ontario
Rexdale (Ontario)

Adresses de l'administration centrale et des bureaux régionaux d'Environnement Canada

Administration centrale
351, boulevard Saint-Joseph
Place Vincent-Massey
Hull (Québec)
K1A 0H3

Région de l'Ontario
4905 rue Dufferin, 2e étage
Downsview (Ontario)
M3H 5T4

Région de l'Atlantique
15e étage, Queen Square
45, promenade Alderney
Dartmouth (Nouvelle-Écosse)
B2Y 2N6

Région des Prairies et du Nord
Twin Atria No. 2, pièce 210
4999-98e avenue
Edmonton (Alberta)
T6B 2X3

Région du Québec
105 rue McGill, 8e étage
Montréal (Québec)
H2Y 2E7

Région du Pacifique et du Yukon*
224, rue Esplanade ouest
North Vancouver (Colombie-Britannique)
V7M 3H7

---

* Un programme informatique pour le calcul de la CL50 existe et peut être reproduit sur une disquette compatible IBM, fournie par l'utilisateur. Pour se procurer ce programme, communiquer avec le Laboratoire de toxicité aquatique à cette adresse.